034929

Coastal Man.

Coastal management

Proceedings of the conference organized by the Maritime Engineering Board of the Institution of Civil Engineers and held in Bournemouth on 9-11 May 1989

Thomas Telford, London

Conference organized by the Maritime Engineering Board of the Institution of Civil Engineers and co-sponsored by the Ministry of Agriculture, Fisheries and Food

Organizing Committee: M. G. Barrett (Chairman), Dr S. W. Huntington, M. J. Wakelin, I. R. Whittle

British Library Cataloguing in Publication Data
    Coastal management: proceedings of the conference organized by the Institution of Civil Engineers and held in Bournemouth on 9-11 May 1989.
    1. Coastal regions. Environment. Management
    I. Institution of Civil Engineers
    333. 78'4

ISBN: 0 7277 1502 X

First published 1989

© Institution of Civil Engineers, 1989, unless otherwise stated.

All rights, including translation, reserved. Except for fair copying, no part of this publication may be reproduced, stored in a retrieval system, or transmitted in any form or by any means electronic, mechanical, photocopying, recording or otherwise, without the prior written permission of the publisher. Requests should be directed to the Publications Manager, Thomas Telford Ltd, Thomas Telford House, 1 Heron Quay, London E14 9XF.

Papers or other contributions and the statements made or opinions expressed therein are published on the understanding that the author of the contribution is solely responsible for the opinions expressed in it and that its publication does not necessarily imply that such statements and or opinions are or reflect the views or opinions of the ICE Council or ICE committees.

Published for the Institution of Civil Engineers by Thomas Telford Ltd, Thomas Telford House, 1 Heron Quay, London E14 9XF.
Printed and bound in Great Britain by Billing and Sons Ltd, Worcester.

Contents

Legislation and policy
1. What is coastal management?   M. G. BARRETT — 1
2. Legislation and policy.   J. R. PARK — 11
3. Technical overview.   I. R. WHITTLE — 21

Coastal planning
4. A county council's approach.   P. BELL — 31
5. Physical constraints on the planning of coastal development.   D. BROOK — 43
6. A regional strategy.   B. HALL — 51

Research
7. Research on tides, surges and waves.   G. ALCOCK — 59
8. Research on beaches and coastal structures.   M. W. OWEN — 77

Caring for the environment
9. Engineering conservation.   A. G. ROBERTS — 93
10. Nature conservation in coastal management.   K. L. DUFF — 105
11. The Australian experience.   I. R. W. MILLER and P. A. MUMFORD — 115

Coastal studies
12. The deterioration of a coastline.   M. J. WAKELIN — 135
13. The Anglian Sea Defence Management Study.   C. A. FLEMING — 153
14. Implications of climatic change.   K. M. CLAYTON — 165

Coastal defence methods

15.	Beach management. K. J. SHAVE	177
16.	Marine resources. F. PARRISH	187
17.	Management of coastal cliffs. A. McGOWN and L. K. R. WOODROW	197
18.	Coastal structures. N. PALLETT and S. W. YOUNG	211
19.	Scheme worthwhileness. E. C. PENNING-ROWSELL, A. COKER, A. N'JAI, D. J. PARKER and S. M. TUNSTALL	227

Engineering studies

20.	Case study at Carmarthen Bay. P. C. BARBER and R. P. THOMAS	243
21.	Evolution of the Bournemouth defences. R. E. L. LELLIOTT	263
22.	Workington ironworks reclamation. B. EMPSALL	279
23.	Wirral scheme. C. D. DAVIES	293

1. What is coastal management?

M. G. BARRETT, FICE, MConsE, Partner, Posford Duvivier

SYNOPSIS. Control of the coastline and coastal activities lies with many different authorities and regulatory bodies. There is also a very considerable variety of interests both public and private involving protection, development, usage and conservation.
Conflicts must be identified and resolved. There is a need for more coordinated long term planning particularly in relation to the coastal sediment budget and to liaison between planners and engineers.

INTRODUCTION.
1. What is coastal management? To a coastal engineer it may mean the planning, construction, monitoring and maintenance of works of coastal defence, while to a planner it may mean the control of development and usage of the coastal zone. To a conservationist it undoubtedly has a further meaning which may well include opposition to the construction of coastal works.
2. The aim of the conference is to bring together members of the various professions and disciplines involved in the management of the coastline in its wider sense. For this purpose the objective of coastal management is defined as the preservation of coastal resources, while at the same time, satisfying the sometimes conflicting requirements of protection, development, usage and conservation.
3. The question of water quality and pollution control, while undoubtedly coming within this definition, has been excluded because of the need to limit the diversity of subject matter. It will in any case be addressed later by a separate conference.
4. The main focus of the invited papers is on the protection of the coastline, coastal planning and environmental impact, economic justification, coastal resources, protection methods and case histories.

THE COASTAL ZONE.
5. In order to discuss coastal management one must first define the coastal zone. There is no legal definition but various bodies have suggested their own in order to define the extent of their particular interests.

6. In addition to the foreshore itself I would include those areas of adjoining sea bed where the processes of sediment transport are linked in some way to the sediment movements of the shoreline or where the activities of construction or industry can affect the shoreline regime.

7. Landward of the shoreline I would include those areas of land likely to be affected in the foreseeable future either directly or indirectly by the erosion of the coast together with areas potentially at risk from inundation by the sea. To this one should perhaps add those areas adjoining the sea where development and usage are directly linked to the activities of the coast.

COASTAL RESOURCES

8. What are the coastal resources which are to be preserved? The list of priorities is usually headed by property including residential, public, commercial and industrial structures together with the infrastructure which supports them. Because of the investment involved property usually commands a high level of financial resource for its preservation. At the other end of the scale is coastal farmland and open country which, in most cases because of its low value, is uneconomic to protect.

9. Other principal resources include:-
   (a) the beaches and backshore areas which provide an amenity for recreation and tourism,
   (b) access to the sea for recreation via the beaches, marinas, etc.,
   (c) industrial and commercial access to the sea via harbours, entrance channels, etc.,
   (d) sea fisheries,
   (e) coastal scenery,
   (f) coastal habitats for flora and fauna,
   (g) geological exposures,
   (h) foreshore and sea bed minerals,
   (i) supplies of cooling water for industrial installations.

CONTROL OF THE COASTAL ZONE

10. Control of the coastal zone arises primarily through legislation, funding and policies of regulatory bodies. There are many Acts of Parliament which relate directly and indirectly to coastal management and no one body has overall responsibility. In the past the principal Acts have probably been those for Coast Protection, Land Drainage (sea defences) and Town and Country Planning, with further legislation relating, interalia, to harbours, navigation, fisheries, pollution and dumping control, highways, mining and dredging.

11. The issues of conservation of sites of scientific interest and the environmental impact of coastal works have become increasingly a matter of public concern leading to the recent introduction of significant laws, statutes and regulations in that field.

12. The application of legislation does of course depend upon the policies of the government departments and the various regulatory bodies. More importantly, it depends upon the grant-aid and other funding associated with those policies for, at the end of the day,

little can be done without money. While profit can be made from development on the coastline, there is unfortunately no profit in monetary terms from works of coastal defence or from conservation.

NATIONAL OR LOCAL CONTROL?

13. Many problems and anomalies have been created in the past, primarily from fragmented control and, to a lesser extent from failure to consider the downdrift effect of coastal works. As a result there has, in recent years, been a call from some quarters for coastal defence works to be the responsibility of a single authority.

14. No coastal engineer would disagree that future works must be undertaken in a coordinated manner with due regard for long term planning and the continuity of coastal processes. Costs must be clearly weighed against benefits in considering consequential effects of works on adjoining coastlines. I do not, however, believe that coastal interests would be best served by total unitary control.

15. The design of coastal defences is one of the few areas of civil engineering in which there are no standardised solutions and where there are still major opportunities for innovative design. Indeed, despite considerable ongoing advances in coastal sciences in recent years, experience still plays a major role in such design. Unified responsibility could well stifle this diversity of approach and lead to a stagnation in design.

16. More importantly, a unified approach might reduce the opportunity to vary designs to suit the inevitably site specific nature of the requirements for each frontage. One feature of coastal defence is the very wide range of solutions available to solve a single problem. This gives the opportunity to take account of local requirements which might not otherwise be considered.

17. The transfer of the administration of the Coast Protection Act 1949 in England to the Ministry of Agriculture Fisheries and Food in 1985 was a major step in harmonising coastal policies. I believe that this gives a sufficient degree of overall control without depriving those authorities responsible for promoting coast protection and sea defences of a degree of local control and choice.

COASTAL PROCESSES AND THE INTERVENTION OF MAN.

18. While man has been able to exercise some limited control over shoreline processes, the general control of wind, waves, tides, surges, offshore sediment movements, and sea and land level changes are still beyond him. It is an essential part of coastal management to fully understand these processes which not only operate along the shoreline but are also inextricably linked to sediment movements and profile changes of the adjoining seabed.

19. Of the various processes perhaps of the greatest importance is an appreciation of the coastal sediment budget. That is to say the source and nature of mobile beach sediments, their pattern and rate of movement, their estimated volume at any point of the coastline and their eventual destination (sink).

20. The measurement of shoreline and nearshore sediment movement still has major shortcomings and some methods of assessment give results which are in error by an order of magnitude. I believe this to be one of the most fruitful areas for future research.

21. The historical effect of man's intervention in the shoreline processes has been considerable, particularly over the last century. The most significant effects have resulted from harbour construction and coastal defence. In many instances the former has caused a major interruption of littoral drift both by the construction of massive entrance piers and, to a lesser extent, by the maintenance of deep access channels.

22. The effect of coastal defences has perhaps been more insidious. The selective protection of developed areas has in many places permanently changed the general alignment of the coastline. The pattern of beach sediment movement has also been extensively altered by:-

- (a) the protection of areas which produce beach material,
- (b) the interruption of littoral movement with groynes and other structures,
- (c) the prevention of natural recession of coastal banks by overwash.

23. Examples of the first are given later and include material both from eroding cliffs and from major natural accumulations of shingle such as Dungeness which have been sterilized by development. In the third group Spurn Point is a classic example of a coastal spit which has been artificially prevented from retreating westwards by overwash to correspond to the rapid erosion of the adjoining coastline.

24. I am not intending to imply that changes resulting from man's intervention are necessarily bad. The Holderness coastline is for instance a good example of isolated protection of urban areas in the past which is compatible with the current concept of re-shaping the coastline by the creation of 'hard points' in the future.

25. One thing is certain, the clock cannot be put back so that future work must utilize to best advantage and accept the consequences of the structures that have been built in the past.

26. There remains an urgent need both for on-going basic research and for both local and regional investigations into coastal processes so that future works harness rather than interrupt these processes. Funding limitations inevitably require that such investigations and research are directed in a cost-effective manner with a clear idea as to how the information gained is to be used in practice. It is important to ensure that the results of research funded with public money are published for the benefit of industry. This has not always been done in the past.

CONFLICTS IN COASTAL MANAGEMENT

27. I have already mentioned the classic conflict of littoral drift interruption both by harbours and other coastal structures exacerbated by the removal of material from the sediment budget as a result of dredging inshore navigation channels.

28. A more recent area of conflict has arisen as a result of environmental concerns. A number of Sites of Special Scientific Interest (S.S.S.I.'s) relate to geological outcrops in cliffs which rely on continuing erosion to maintain fresh exposures. Any form of protection designed to halt or even to reduce the rate of erosion is likely to result in the exposure being eventually buried in colluvium.

29. In many cases coast protection schemes have been required to exclude certain frontages which were subject to this restriction. I am also aware of at least one instance where the owner of a cliff top property was refused permission to protect his own frontage where the outcrop was regarded as unique. Sould the question of compensation be addressed in such cases?

30. There is a further more vital area of potential conflict which I believe requires very careful consideration in relation to long term planning. This is the issue as to whether sections of the coastline should be dedicated to sediment supply by ensuring that erosion is allowed to continue for the benefit of other frontages.

31. Whilst in many cases there is economic justification for augmenting the natural supply of littoral sediment with artificial nourishment, the cost of achieving on a very wide basis in the absence of any natural supply would be inconceivable, even assuming that the necessary resource was available.

32. A classic example of a lost source is the Selsey peninsula in West Sussex which stands at the western extremity of the Sussex/Kent South Coast littoral system. At one time, before the construction of the various harbours on this coast, some of the flint shingle derived from Selsey probably eventually found its way as far as Pegwell Bay despite intermediate accretions at Dungeness and the Crumbles.

33. The Selsey peninsula is rich in gravel and historically eroded at an average of 1 metre per annum over the whole frontage with local rates of up to 9 m. per annum. It is now fully protected although at the present time fortunately receives a supply of some 5,000 cu.m per annum of shingle from offshore under a mechanism associated with weed-rafting.

34. An example on the other hand of a continuing source is Sandown Bay in the Isle of Wight where the beaches are supplied by the erosion of the massive Lower Greensand cliffs at the south end of the bay. In this instance the problem of the consequences of protecting the cliffs at the present time does not arise because it would not in any case be practicable.

35. There are, however, many examples where erosion provides an on-going supply of shingle and/or sand and where, if progressive protection continues, the beaches, both locally and downdrift, are likely to suffer further in the future.

36. Should such areas be subject now to long term decisions so that owners, planners and engineers know where they stand or should we rely on future technological developments to solve these problems when they arise? On the same theme should further development be prevented at those places on the south and east coasts where there are still massive accumulations of beach shingle?

37. There are many other areas of lesser conflict particularly

LEGISLATION AND POLICY

between conservation and usage, such as that between dune control and public access. Conflicts can also arise between amenity and methods of protection. For instance structures using rock or concrete armour units are often unacceptable on grounds of amenity or safety.

LIAISON BETWEEN PLANNERS AND ENGINEERS.

38. I have for some time thought that there should be closer liaison between planners and engineers in the fields of coastal development and protection. Perhaps this conference is a good forum to air these views.

39. Apart from cases where coastal works are specifically provided to protect some new development, they are usually designed to provide sufficient protection for the frontage as it exists. Indeed the economic appraisal for the proposed works is usually required to be undertaken on the current usage.

40. For frontages which are undeveloped this may, for instance, lead to an acceptance of limited overtopping under extreme environmental conditions because of the need to relate cost to benefit. At some later date such a site may be seen as being protected and suitable for development without any appreciation of the design standards originally adopted. Such a sequence of events has, in the past, lead to major damage of property.

41. In one instance I had occasion to design a sea wall which, over a part of its length, protected low-lying ground. A flood bank located well landward of the sea wall served to contain any overtopping which might occur under extreme conditions and prevent a repetition of flooding of an adjoining estate. Imagine my consternation on returning a year or so later to find the area developed and the bank partly removed. It had been assumed that, with a new wall, the bank was redundant.

42. A further example of the problems which arise in connection with coastal planning is the situation in which development is permitted on an unprotected coastline and where the life of the property is limited by on-going erosion. In many instances costly protection has subsequently been necessary to save property which has been built too close to the sea.

43. More than once I have suggested at particular locations that a development limit should be established at a reasonable distance away from an eroding cliff in order to avoid this. The response has been that this could lead to compensation because an owner was entitled to make use of a cliff top plot over its residual life.

44. On one occasion a local authority client refused planning permission for a cliff top bungalow on the grounds that the estimated residual life of the plot was very limited. However, the development was allowed on appeal. The problem here is that the original owner can then sell and the limitation on the life of the property may not be revealed by searches. The new owner may then expect the coast protection authority to take some action to protect it. I recall meeting a pensioner who had just spent his life's savings buying a cliff top cottage which the coast protection authority had, only a year before, decided could not be economically protected.

45. A further variant on this theme involved the construction of an estate of holiday apartments at the crest of a high clay coastal slope which had previously been protected and stabilised by deep drainage. An old war department sea wall at the foot of the slope had unfortunately been allowed to collapse and, as a result of toe erosion, the slope had been destabilized and was well on the move. Planning permission had been given and the engineer for the developer had produced calculations showing the slope to be stable but only in the absence of erosion.

46. Because the land, on which many of these properties were to be sited, was likely to be affected by landslip generated by the erosion, local authority approval under building regulations was refused. The case was taken to appeal and won by the developer on the grounds that building regulations only required the sub-soil to be suitable for foundations at the time of construction.

47. Fortunately the developer at least agreed to omit the seaward row of buildings. However, future owners of the next row will doubtless have a surprise when the limitation on the residual life of their property becomes apparent to them.

HARD VERSUS SOFT

48. Much is made these days of the merits of what is generally referred to as 'soft' as opposed to 'hard' defences as if it was something new.

49. I suppose that a truly 'soft' defence is one comprising artificially nourished foreshores where the beach is allowed to take up a natural alignment totally unfettered by any fixed structure. Such schemes are few in number and indeed most schemes these days involve the rehabilitation of existing old 'hard' defences.

50. So called 'hard' defences have suffered from the reputation derived from the massive vertical structures erected by our Victorian forbears and subsequent generations, largely to accommodate promenades at major seaside resorts, in ignorance of long term consequences. The well known classic example of this is the New Brighton sea wall on the Wirral. This is a massive vertical structure sited well seaward of the shoreline on a foreshore with fine sediment, high tidal range and strong alongshore currents.

51. Nothing is ever entirely black or white and it can be dangerous to categorize coastal solutions too rigidly. In general sea walls are (or should be) designed in conjunction with proper considerations of beach management so that they become a combination of 'hard' and 'soft'. In this context the 'hard' element of such a scheme may have the following advantages:-
    (a) it can provide 'fail-safe' protection if the beach is not maintained,
    (b) it can provide protection against the consequences of extreme events such as overtopping and draw-down,
    (c) it may enable considerable economies to be made in the volume of beach material to be supplied and maintained,
    (d) it often allows secondary considerations such as access and amenity to be met.

LEGISLATION AND POLICY

52. I have certainly used beach nourishment regularly for some 30 years for beach management in conjunction with sea walls. It is only more recently, of course, that it has been adopted in this country on a more massive scale.

53. There are unfortunately many examples of schemes involving hard defences which have not, in the long term, performed well. In some instances this has been due to a failure on the part of the designer to anticipate the effect of the sea wall on the foreshore or to properly consider future strategy for beach management.

54. Very often, however, it has been due to a failure to maintain the works originally provided. This includes allowing groynes to fall into disrepair or failing to provide on-going beach recharge either as a result of a policy change or, more frequently, economic restrictions.

55. In reviewing the effect of failing to maintain beaches in front of certain hard defences, one might perhaps pose the question as to what the consequences might have been had the original schemes been totally reliant on beach recharge.

FUTURE RESOURCES FOR BEACH NOURISHMENT

56. I have no doubt that there is a consensus of view amongst coastal engineers that the ideal form of coastal defence in purely engineering terms is a massive beach. Whether this can be achieved in future economic terms and in proper long term use of available resources is quite another matter.

57. Economic appraisal of a scheme which depends primarily on beach nourishment requires an accurate assessment of the future movements of that material and long term estimates both of the future cost and availability of recharge material. Reliable estimates of future movement have in some cases been found difficult to make and should therefore be subject to a sensitivity analysis in any economic appraisal.

58. Future material resources for massive beach recharge schemes are of vital importance both in terms of type and quantity of material available. Availability also obviously affects cost.

59. With the progressive protection of the coastline and sterilization of natural accumulations of shingle by coastal development, natural sources of material are becoming rapidly reduced. The demand for beach nourishment may be expected to increase substantially as natural supplies decrease. At the same time there are other increasing demands on potential nourishment material, such as offshore dredged gravel for the aggregate industry, a resource that is ultimately finite.

60. In order to meet such a requirement I believe that it will be necessary firstly to compromise on specifications in order to use material that is available and which does not compete with other demands. Secondly it will be necessary to avoid extravagant use of quality material.

61. The unrivalled durability of the flint shingle that occurs both along the shoreline and offshore in south-east England makes it an ideal nourishment material. Eventually it may become necessary to consider compromise alternatives such as crushed stone, as indeed has already been used in limited quantities in other parts of the country and been subjected to durability tests.

SEA LEVEL CHANGE

62. Another important matter of long term coastal planning and one which is addressed by one of the conference papers is the question of secular sea level change relative to the land.

63. The significance and possible consequences of such change are fortunately now a matter of international and public awareness and concern. However, views expressed as to the extent of such change vary in some instances by an order of magnitude.

64. There is clearly some considerable urgency for establishing more accurate predictions as soon as the necessary research enables it to be done. This should suppress some of the more alarmist views and introduce a more coherent policy for long term design.

ALTERNATIVE SOLUTIONS

65. A topic which is repeatedly raised in coastal circles as a result of economic restrictions on public expenditure is the question of low cost coast protection. In some quarters there appears to be the view that there is some, as yet undiscovered, cheaper method of coastal defence.

66. All schemes which are grant-aided are subject to financial scrutiny and economic appraisal and should therefore cost no more than is necessary to achieve their fitness for purpose.

67. While clearly research must continue into cheaper materials and less costly construction methods it would be foolish to imagine that low cost coast protection is an option other than for short term solutions.

CONCLUSIONS

68. There is clearly considerable scope for better long term planning and for closer liaison between engineers, planners, conservationists, developers and the various regulatory bodies in the context of coastal management.

69. This will involve both general research and local investigations into the coastal process and considerations of the consequences of long term change in order to provide a consistent future policy.

2. Legislation and policy

J. R. PARK, BSc, PhD, FRIC, FIFST, Head, Flood Defence and Land Sales Division, Ministry of Agriculture, Fisheries and Food

SYNOPSIS. The two Acts of most immediate relevance here are the Land Drainage Act 1976 and the Coast Protection Act 1949. The first of these, in this context, is concerned with protection against sea and tidal flooding in low lying areas; the second gives power to carry out works to prevent permanent loss of land by erosion and encroachment by the sea. The paper outlines some of the history leading to the present legislation; notes the way the Acts are applied to different parts of the coast; explains some of the administrative requirements underlying approval of schemes; and gives an indication of the system of exchequer grant support and capital control.

INTRODUCTION

1. Currently there are two main statutes conferring powers to carry out work to protect the coastline in England and Wales. Protection against flooding of low lying areas (in this context usually referred to as sea defence) may be carried out under the Land Drainage Act 1976, and works to prevent erosion of land and encroachment by the sea may be undertaken under the Coast Protection Act 1949. Currently Water Authorities primarily carry out sea defence work and their powers will transfer to the new National Rivers Authority. Maritime District Councils nevertheless also have such powers as do Internal Drainage Boards, although they rarely use them. It is possible for other agencies to carry out sea defence work.
Coast protection work is usually undertaken by Maritime District Councils, who are designated Coast Protection Authorities under the Act. There is provision for other parties to do such work, with the Coast Protection Authority's consent. In fact, it is generally expected that large undertakings (such as public utilities for example) should carry out their own coast protection work under these provisions. For each Act the powers are permissive and blanket protective duties are not imposed.

LEGISLATION AND POLICY

HISTORY
2. It might be useful to sketch the history of sea defence and coast protection arrangements.

Sea Defence
Some responsibility for protection against flooding has been within the remit of land drainage authorities for centuries. Links with Central Government can be traced back as far as Henry VI when Commissions of Sewers were required to report on flooding problems throughout the land. In parts of England it is possible to detect a series of sea walls that have been built over the years, showing the long history of attempts to keep the sea at bay. The more up-to-date picture probably begins with the Royal Commission appointed in 1927 to review land drainage. A result of the deliberation was the first of the modern statutes in this area, namely the Land Drainage Act 1930. This Act included defence against sea water within the definition of drainage. It set up 49 Catchment Boards to tackle the more urgent problems arising from previous neglect of sea defences. But the Boards were somewhat restricted in the work they could do as they were only allowed to operate in connection with main rivers (These being designated as such on Ministerially approved statutory maps). Unfortunately the Second World War intervened and defences fell back into neglect.

3. Relatively minor change was introduced by the River Boards Act 1948 which reorganised the Catchment Boards into 34 River Boards and it was they who suffered the consequences of the earlier neglect when the North Sea storm surge occurred in 1953. The East Coast was seriously damaged by tides of unprecedented levels and it is fair to say that even well maintained defences would have been overwhelmed. After this disaster, an inter-departmental committee was set up under the chairmanship of the Viscount Waverley to report, amongst other things, on the causes of the surge and the possibility of a recurrence. One of the recommendations of the committee was that sea defences should be improved to withstand surge tides of 1953 proportions and it was no coincidence that the Land Drainage Act 1961 extended River Boards powers to construct sea defence works wherever the need arose. Waverley also recommended that flood warning systems should be set up so that early action could be taken in advance of any severe future events. The improved defences and the warning systems (there are separate systems for the east coast and west coast) have since proved their worth on numerous occasions, including instances where the 1953 levels have been exceeded.

4. In 1963 the process of rationalising the organisation of drainage authorities took a further step. This time, the Water Resource Act 1963 reduced the number of River Boards

to 27 and called them River Authorities. Also, in what proved to be a first step towards unification of water management, the Act gave the River Authorities additional responsibility for water resources.

5. The unification process reached its apex when the Water Act 1973 introduced the concept of totally integrated river basin management. Ten Regional Water Authorities were established in England and Wales to oversee all water supply, sewerage, river management and regulatory functions, including responsibility for flood and sea defence. However the authorities were required to discharge their land drainage, flood and sea defence responsibilities through Regional Land Drainage Committees and, in many areas, subordinate Local Committees remained mirroring the River Authority areas. As previously, land drainage and flood defence was funded by precept on County Councils and in recognition of this the councils were given majority membership of both committees.

6. And now, of course, the Water Bill is currently in Parliament. This provides for the privatisation of the water and sewerage utilities and sets up a National Rivers Authority (NRA) to oversee river management and regulation, including a strong environmental responsibility. Flood defence will therefore become the responsibility of the NRA but as with Water Authorities the functions will be discharged through regional and local committees. However the committees are being re-named 'Flood Defence' committees to more aptly describe their primary function.

7. In the meantime a composite Bill had been introduced to update the land drainage statute. This became the Land Drainage Act 1976 and although it has been subject to various changes since, it still sets out the legislative background, powers and responsibilities concerning sea defence.

Coast Protection

8. Turning to coast protection, Local Authorities had no general statutory powers to construct works before 1949. However, many of them had provided defences under general powers such as those applying to parks, pleasure grounds and open spaces. In addition, some authorities managed to obtain special powers under local Acts. But the first national consideration of coastal erosion began with the Royal Commission appointed in 1906. It recommended, amongst other things, that there should be controls on the removal of beach material and that local authorities' works at the shore should be brought under central supervision. But the Commission saw no case for making coast protection a central Government responsibility. Little came immediately from these recommendations, until the Coast Protection Act 1939 introduced provisions for the control of the removal of beach material. The Board of Trade was empowered to make

LEGISLATION AND POLICY

orders restricting the removal of such material wherever there was thought to be danger of erosion if the beach was reduced. These powers in effect extended similar legislation dating back to Harbour Acts of 1814 which empowered the Admiralty to prohibit the removal of shingle and ballast from the shores or banks of any "port, harbour or haven".

9. Here too, protective structures deteriorated during the Second World War. Following severe storms in 1946, investigations showed that maintenance of coast protection works had been neglected and that great erosive damage had occurred. It was recognised that private owners lacked the funds necessary for technical advice or to carry out the extensive works necessary. It was therefore considered that some measure of co-ordination should be introduced for the necessary work to be undertaken. The Coast Protection Act 1949 thus gave Local Authorities powers to carry out works under general supervision by central Government. This latter provision has helped to ensure that works are effectively designed and, in particular, that there is at least some regard to effects further along the coast. The 1939 Act was repealed, and its provisions concerning removal of beach material were continued under the new Act.

APPLICATION OF THE ACTS TO DIFFERENT PARTS OF THE COAST

10. The Coast Protection Act sensibly defines the sea as including rivers to their tidal limits with the exception of certain harbours and estuaries specified in the 4th Schedule. Thus it is clear that the Act applies, in principle, to all stretches of coast within the areas of the Coast Protection Authorities. "Protection" is defined as "protection against erosion or encroachment by the sea" and, left unchecked, the sea of course eventually takes over causing the distinction between coast protection and sea defence to enter a grey area - particularly where higher land descends to sea level. In, fortunately, rare circumstances there may be doubt whether the Water Authority or the Maritime District Council carries the responsibility for remedial works.

11. This point was examined by the Waverley Committee in the course of its investigation into the causes of the 1953 floods. The Committee reviewed the financial and administrative functions of bodies with powers to construct and maintain sea defence and coast protection structures and saw some apparent advantage in bringing both types of coastal defences within a single system supervised by one type of authority. But it recognised that the River Boards (as they then were) and Coast Protection Authorities had different functions to perform and concluded that there were no stretches of coast where interests were duplicated or left completely unconsidered by either type of body. Nevertheless it was recommended that the two authorities

should co-operate closely by agreeing between themselves (or, in cases of dispute, with the help of central Departments) arrangements for exercise of their statutory powers within their respective areas. Thus, the functions of the Coast Protection Authorities and Water Authorities (as representative of organisations undertaking sea defence work) are seen to be distinct although the two types of authority liaise where necessary. In fact, the Coast Protection Act goes a little further than this in that it provides for adjacent Maritime District Councils to establish Coast Protection Boards which would include the appropriate Water Authority. In this way the Water Authority would be able to exercise some responsibility for co-ordination of coast protection works. However, no such boards have been established. Nevertheless, the Coast Protection Act requires a Coast Protection Authority to obtain consent before carrying out any activity on works constructed or maintained by a Water Authority. There is no reciprical provision under the Land Drainage Act 1976.

12. Given that sea defence and coast protection were defined to be different and that there was facility for liaison between the various bodies involved, it seemed for a long time that any confusion over responsibilities would be minimal and of no consequence. However, the picture changed slightly after 1978 when Canterbury City Council proposed a coast protection scheme at Whitstable. The main purpose of the scheme seemed, in the view of the Department of the Environment, (who then administered the Act), to be to prevent flooding, and as such was more appropriate to the Land Drainage Act. Canterbury did not accept this and eventually the disagreement reached the High Court. The proposed scheme consisted of a sea wall 22 feet high which protected against erosion as well as flooding. The judgement of the Court was that since the interactions of erosion, encroachment and flooding are often inseparable, prevention of erosion could itself prevent flooding. Thus it was decided that the wall could be considered to be a coast protection structure. Since this case, it has become clear that where works prevent both flooding and erosion, the work can be carried out under the Coast Protection Act.

13. The so called 'Whitstable Judgement' led to the review of separate coast protection and sea defence schemes in the Green Paper on Financing and Administration of Land Drainage, Flood Prevention and Coast Protection issued by MAFF in March 1985. This proposed integrating arrangements for coast protection and sea defence under a new category of "coastal works". Water Authorities would have held overall general responsibility, but Maritime District Councils could have been designated as the Coastal Works Authority over particular lengths of coastline in their areas. In practice it was anticipated that such designation would mainly cover important amenities such as seaside resorts. The Maritime

LEGISLATION AND POLICY

District Councils would have funded the schemes and would have been eligible for MAFF capital grant in the normal way, although some rationalisation of grant rates paid to water and local authorities was implicit. The Water Authority would have needed to be satisfied that Councils coastal defence programmes had no adverse effects on the adjacent coastline, and would have had powers to carry out works it judged necessary when the Council had not acted.

14. As a first step towards integration, responsibility for the Coast Protection Act was transferred from DOE to MAFF and it is now administered in harness with the Land Drainage Act 1976. The maximum rate of capital grant for coast protection was reduced from 79% to 70%, but it is still higher than rates of grant for sea defence. The matter has progressed no further, mainly due to lack of opportunity for the necessary primary legislation. The Green Paper's proposal attracted a good deal of hostility from the Maritime District Councils, who wished to retain control over their coastlines. County Councils also argued that strategic control of coastal policy was within their province but Water Authorities were generally attracted to the Green Paper proposal. There the matter rests, at least for now. Integration would require further consideration of complex arrangements and some difficult financial and responsibility issues would need to be resolved. Preparatory work is unlikely to be put in hand until a clear opportunity emerges for the necessary legislation.

ADMINISTRATION OF COAST PROTECTION AND SEA DEFENCE SCHEMES

15. When a Coast Protection Authority wishes to carry out a capital scheme the Act requires that a notice be published in local newspapers and (if the scheme exceeds a cost of £50,000) the London Gazette. The authority also has to notify certain other interested parties, such as their neighbouring authorities and the Water Authority, and objections may be lodged. Where objectors consider the proposed works detrimental to the protection of land specified in the notice or believe they interfere with their functions under any other Act, a public local inquiry may be held. In all other cases the proposed scheme is examined by MAFF engineers who may allow other types of objectors to state their case informally. In either circumstance, the Inspector at an inquiry or the engineer makes recommendations to the Minister. The Minister is required either to approve the scheme as submitted or with modifications, or to direct the authority not to carry out the work.

16. In circumstances where a Coast Protection Authority has to carry out work as a matter of urgency, the requirements for publishing notices etc are waived. As soon as possible, however, the authority is required to notify

the Ministry and the Water Authority that such work will be, or has been, commenced.

17. The Land Drainage Act does not require that notice of sea defence works be published. The system of objecting to proposed schemes is less formal and work can be carried out without Government approval. However, most schemes are carried out with MAFF grant aid for which specific approval is required. In contrast, grant aid approval for coast protection schemes is, in effect, linked to approval of the scheme itself.

18. Three main criteria are taken into consideration in the approval of both types of schemes; these are technical soundness, economic justification and environmental sensitivity. It is fundamental to ensure that a scheme is technically appropriate and will do the job asked of it throughout its design life (with adequate maintenance of course). Expert advice is available from MAFF engineering staff. Similarly MAFF checks to ensure that the benefits of a scheme are at least equal to the costs. In practice, environmental sensitivity has been assessed in similar ways for both Acts, although there have been procedural differences and the statutory practice is now very different.

19. Promoters of sea defence works, be they Water Authorities, Local Authorities or Internal Drainage Boards, have responsibilities under Section 48 of the Wildlife and Countryside Act 1981 to further conservation and take environmental effects of works into account insofar as is consistent with their responsibilities under the Land Draiange Act; this responsibility has been discharged by consulting the Nature Conservancy Council when drawing up proposals for works under joint guidelines issued by MAFF/DOE/WO. The Act has proved successful in significantly reducing conflict between land drainage and environmental interests and in recent years it has been rare indeed for MAFF to receive grant aid applications for schemes causing environmental concern. Section 48 does not however apply to coast protection works and MAFF has therefore itself consulted NCC and other appropriate environmental agencies. Section 48 is shortly to be superseded by Section 7 of the Water Bill.

20. A further difference emerged with introduction of the EC Environmental Assessment Directive under Regulations covering England and Wales in July 1988. In general an environmental assessment is required where proposed works are considered to have a significant environmental effect. For the most part, planning authorities will decide whether environmental assessment of proposals is appropriate under DOE guidelines and will consider the outcome when deciding upon planning consent. New flood defence works require planning consent and are considered accordingly. Improvements to flood defence works have deemed consent

LEGISLATION AND POLICY

under the General Development Order and are therefore subject to special procedures under the Environmental Assessment (Land Drainage Improvement Works) Order 1988. These require promoting authorities to announce proposed works in the local press, say whether or not EA is proposed and then only proceed with the works if no 'environmental' objections are received, or if they are resolved, possibly after appeal to the Minister. When EA is carried out its results are also announced in the local press and any sustained objections to the works are referred to the Minister for decision. In the few months of operation no difficulties have occurred.

FINANCING

21. Generally, Water Authorities fund sea defence and all other land drainage costs by precept on the County Councils in the authority's area and these are met from the general rates; MAFF pays capital grant on approved works. Local Authorities also receive Government grant for approved sea defence or coast protection schemes with the remainder funded from normal sources of local authority income.

22. Grant rates for coast protection schemes are calculated from a formula taking account of the notional burden a scheme places on the local rates, the existing rate burden from previous approved schemes and the authority's rate poundage compared to the National average. Provided that the result of this calculation is above a certain threshold then grant is payable. At present, the resultant grant rates vary from 24 per cent to 70 per cent of a scheme's capital cost. For sea defence schemes, grant rates are fixed for Local Authorities at 41 per cent, including a supplement of 15 per cent for tidal works. Grant rates for Water Authorities are fixed for individual land drainage districts and currently range from 30 per cent to a maximum of 65 per cent, including tidal supplement of no more than 15 per cent. In England, total grant provision in 1988/89 was about £10.7m for sea defence and £13.4m for coast protection. For 1989/90 the figure for sea defence grant will be about £11m whilst that for coast protection will rise to about £15.1m.

23. The grant and capital provision for flood defence work is currently increasing. Additional funding will be made available over the next three years: for Water Authorities, the increase will amount to £15.9m grant, and for Local Authorities additions to the grant aided capital programme will produce a grant increase of about £1.8m. In each case about half of the capital funding is likely to be directed to sea defence schemes. For coast protection, there will also be an increase in the capital programme over this period, equivalent to a grant of about £3.9m.

24. For coast protection schemes only, where Government grant is made available, the county containing the Coast Protection Authority is obliged to contribute to the cost of the scheme. The amount of the contribution is for agreement between the County and District Councils. If agreement is not possible the Minister has power under the Act to determine, within limits, the amount of the county's contribution. In practice, counties generally contribute about half the net of grant costs.

25. All other contributions are subtracted from the cost of a scheme before grant is calculated. The Coast Protection Act still has provision for works schemes, which are legal documents specifying proposed work (but otherwise very similar to normal proposals), which allow Coast Protection Authorities to require beneficiaries of schemes to contribute to costs. However those latter powers have been suspended by administrative action and now those benefitting from coast protection work may be invited to contribute voluntarily; the invitation is normally extended only to commercial and industrial interests. The same voluntary practice is applied in the case of sea defence work.

26. There are also controls on capital spending by Water Authorities and Local Authorities. For Water Authorities this control is exercised by way of the Grant Earning Ceiling. This sets an upper limit to the capital spending that can qualify for grant aid, although, if the resources exist, the authorities can still undertake sea defence work without the benefit of grant aid. The authorities submit to the Ministry each year their medium term plans for capital spending. These plans include their arterial drainage as well as sea defence work. They are examined as part of the consideration by Government of all public expenditure. Usually by early winter, the authorities are informed of their individual Grant Earning Ceilings for the following year. If, for any authority, the ceiling is beneath its works programme priority will usually be given to sea defence schemes over other flood defence work. In 1988/89 the total Grant Earning Ceiling for English authorities amounted to £37.6m, excluding Seaford, of which about half is for sea defence work. The equivalent figure for 1989/90 is £42.5m.

27. Capital control on local authorities is by way of a system of capital allocation, which effectively means permission to spend. The Department of the Environment gives block allocations for authorities spending on various services. This can be used as well as additional allocation specifically for coast protection or sea defence work as the case may be. In practice, it is usual for schemes to receive the specific coast protection or sea defence allocation, whilst resources remain, at 100 per cent of requirement. In 1988/89 allocation totalled £20.7m (in England).

LEGISLATION AND POLICY

CONCLUSIONS
28. There are distinct differences between the two Acts but there is no evidence that they cause major difficulties. Administrative practices have been and may continue to be brought into closer alignment as far as possible within the limitation of the statutes. At present though, no early major changes are expected in the way the two Acts are used to help support the construction and renewal of sea defence and coast protection works.

ACKNOWLEDGEMENT
This paper has been prepared with the invaluable support of colleagues in the Division, particularly Mr D Buckle and Mr R Baker.

3. Technical overview

I. R. WHITTLE, MICE, FIWEM, MBIM, Chief Engineer, Flood Defence and Land Sales Division, Ministry of Agriculture, Fisheries and Food

Synopsis

Engineering solutions to coastal problems may be undertaken under the Coast Protection Act 1949 or the Land Drainage Act 1976. Although the designs of the works will vary widely in concept they have to meet many common criteria. Work funded from the public purse will normally be expected to provide an appropriate standard of defence, be environmentally acceptable and worthwhile. Frequently solutions offer the opportunity to enhance the amenities and the environment.

Engineers at the Ministry of Agriculture, Fisheries and Food (MAFF) assess the worth to the nation of these schemes and ensure that the relevant technical criteria are adopted and that latest design and construction techniques are employed whenever possible.

Extensive research and development programmes into the complex range of coastal processes are funded through the Chief Scientists Group, and the results are used to assist with this detailed examination. Now and in the near future the effects of the prediction of sea level rise has to be addressed and suitable policies developed for the UK.

The paper highlights some of the important issues which have to be addressed by coastal engineers in the preparation of schemes for grant aid.

Sea Defences

The most serious storm within living memory to attack the east coast of Britain occurred in January 1953. Many thousands of hectares of land were flooded, thousands of houses and factories were damaged and some 300 lives were lost. It was estimated that some 1200 breaches had to be repaired and countless kilometres of sea defences made good following extensive overtopping along 2000 km of defences.

A departmental committee, under the chairmanship of Viscount Waverley (1), known as the Waverley Committee, was set up to

LEGISLATION AND POLICY

investigate, inter alia, the causes, consider the standards of future defences and the need for a public warning system.

This Committee recommended that the water level (tide plus surge) reached in 1953 should be the general maximum standard of protection which could be reasonably afforded. Authorities embarked upon an extensive programme of reconstruction of defences in compliance with this policy.

Since 1953 a number of storms have been experienced in which levels exceeded that of this earlier storm. Most of the reconstructed defences have withstood these onslaughts although some older defences suffered and failed in a number of areas.

Following on from the Waverley Committee two committees were set up to oversee research programmes. These were replaced by the Flood Protection Research Committee chaired by Sir Angus Paton (2). In its report to the Ministry of Agriculture, Fisheries and Food in 1979 it recommended the adoption of an increase in defence level in recognition of the recent increase in storm frequency and magnitude. The recommendation directed that the effects of waves to be taken into account together with the worst recorded still water levels, thus allowing protection against the worst recorded storm tide. The Committee recognised that there were many combinations of wave/surge/tide which affected the final level at the defence structure and it recommended that each element should be analysed and combined to determine overall risk to the protected areas.

The Committee retained the earlier recommendation that the standard of defence to be provided at public expense had to be commensurate with the value of benefits justified in terms of the general public interest. This did not preclude a beneficiary from making an equitable contribution to secure an even higher standard.

An important phrase retained by the FPR Committee ensures that departments and authorities have the freedom to judge whether to spend public money on the provision of a defence system to protect against an unreasonably severe event. It is of interest to note other recommendations contained in the earlier report. These were concerned with the engineering solutions for the improvement to the defences such as retention of secondary lines of defence, construction of 'cross banks' to contain overspill, the protection of soft embankments to withstand overtopping of water. Many schemes have been devised in which this latter recommendation has been adopted.

Repairs of these damaged defences and reconstruction of new defences were affected by either the construction of a new concrete or steel piled wall or the reconstruction of a clay embankment having the front wall, possibly the crest and back face, protected by concrete blockwork or other semi-rigid material, or reconstruction of sand dunes in combination with other works such as groyne fields to stabilise the foreshore.

Paradoxically whilst many of these schemes solved the problems and prevented sea water gaining access to the hinterland they created new problems on the coast line.

The solid faces, vertical or stepped, to the defences, generated reflected waves which caused scouring of the foreshore and lowered the beach levels thus exposing the toe foundations and endangering the stability of those defences. Much effort has since been expended in constructing works which 'chase the toe downwards'.

Many new defences have been built in front of sand dunes or land containing soft erodable geological formations and these in common with similar coastal works which protect eroding cliffs, prevent the sea from 'winning' material needed to satisfy the natural littoral drift processes.

The littoral drift budget must be satisfied otherwise downdrift beaches can become denuded, thus introducing new hazards in these adjacent stretches of coastline. It is recognised that structures such as major terminal groynes, solid breakwaters or piers, to say harbour entrances and even groyne fields unless filled at construction stage, can all interrupt this drift process and aggravate an erosion problem.

Coast Protection

Defences which protect soft cliff and other vulnerable ground against erosion are not built with the benefit of any national guidelines. However, since they are required to resist similar wave and surge forces created by stormy seas, they can be designed using common criteria. Thus the engineering solutions may be akin to these used in sea defences and so create similar operational problems of lowering or accelerated erosion of adjacent beaches. Not all defences need to be solid and alternative designs may be appropriate. There are many examples of coastal solutions which consist of either permeable structures or perhaps strategically placed, reinforced headlands or strong points. These structures merely slow down the rate of erosion and ensure that littoral drift demands are met in part. In environmental terms such solutions ensure that geologically-interesting landforms are maintained for geological and coastal morphological study.

LEGISLATION AND POLICY

The Environment

One of the principal forces which structures have to resist arises from the energy of momentum in storm waves and surges. Today, coastal design structures are designed to dissipate these high levels of energy in the least-harmful, least-damaging way. Solutions may involve the reconstruction of beaches by artificial beach-feeding programmes, encourage regeneration of saltings, construction of rock-armoured headlands or off-shore islands.

These structures have the added advantage of enhancing the environment, albeit that the off-shore structures may be the subject of criticism about their visual effects in the immediate post-construction period. Experience in this country with off-shore islands indicates that marine life establishes itself on the rock-armour material thereby creating a new and interesting marine environment. In the case of the recharged beach a safe refuge is created for the ardent sea-angler or, after the material has been 'sorted' by the sea it can provide sandy zones to the delight of the younger visitor keen to build flood defences around his newly built fortress!

Most engineering solutions have an impact on the environment and care is needed to ensure that, as far as possible, the solution is in keeping with the adjacent areas. For instance if the geological strata lie horizontally then the solution should attempt to replicate this formation and if possible the choice of armour rock should ensure that there is some degree of harmony with the local geology. Solid concrete structures can be harsh and uninteresting so plain faces should be given some surface treatment. Perhaps a relief detail can be cast into the concrete or some surface treatment can be applied after removal of the shuttering.

Changes to the landscape are inevitable because of the many demands being made upon it by society. Many of these demands are considered damaging to the environment and new legislation has recently been introduced to limit and, where possible, reverse these adverse effects.

The Wildlife and Countryside Act 1981 imposed duties on authorities to ...further the needs of conservation. This Act strengthened the provisions contained in the Water Act 1973 and embraced the activities of drainage authorities. The application of the provisions has been extended so that the spirit of the Act is observed by drainage bodies as defined by the Land Drainage Act 1976.

In 1988 two new Statutory Instruments (SI) (3,4) were introduced in response to a European Council Directive,

85/337/EEC. These SIs set out the provisions for the preparation of Environmental Statements. One SI strengthens the existing planning legislation and the other strengthens the conservation procedure for certain types of land drainage activity.

Cost Benefit Analysis

Schemes of a capital nature which provide protection either against erosion or from flooding by the sea may be eligible for grant aid by MAFF. Under the two Acts referred to above, government may offer authorities financial assistance to meet the high costs of the capital investment.

Proposed schemes meeting, inter alia, technical standards of performance and durability have to offer benefits which can be shown to give a minimum rate of return. This rate is set by Treasury and is currently set at 5%. Economists have argued both that this rate is either too low or too high! Investment in capital schemes yields benefits over a long period of time and it is necessary to discount streams of both costs and benefits to arrive at Present Values and subsequently benefit/cost ratios and Net Present Values.

Clearly this type of analysis introduces subjective assessments and it is necessary to demonstrate the robustness of the initial analysis. The valuation of elements of the scheme or components of the costs-stream or the benefits-stream which are time-dependent are open to question. These valuations must be subjected to sensitivity tests to establish the confidence limits for the worthwhileness of the project.

MAFF have supported the development of these analyses through contracts with the Middlesex Polytechnic Flood Hazard Centre. This Centre now holds extensive data bases of, what has become generally accepted as, standard values for the built environment. Elsewhere, other agencies have been involved in the further development of methodologies and the establishment of data stores to assist with analyses. Many authorities and consulting engineers turn to the Flood Hazard Reearch Centre to gain access to these extensive data banks and to universities and land-use consultants for data relating to the rural environment.

New and innovative components are presently being examined and validated for deriving the benefit assessment. Other countries, for instance America, are further advanced and more relaxed with the assessment of benefits. Some of these principles have been developed further and new ground broken in the use of contingent valuations by the University of East Anglia under a recent contract with Anglian Water

LEGISLATION AND POLICY

Authority to justify a proposed capital scheme. The need for evaluating benefits is really an essential safeguard whereby investment, particularly from the public purse, can be demonstrated to be justified. Many authorities are now adopting this principle to support decisions made by management to justify their very extensive maintenance programmes.

Research and Development

Funds for the research programmes undertaken by MAFF are provided through the Chief Scientist. Currently some £1.2m is being spent annually on research programmes with the NERC Institutes and Hydraulics Research Ltd and a lesser sum is spent on other research by universities under contracts with MAFF.

Results from these programmes are complementary to other national and international programmes which are supporting basic or applied research activities.

Attention is directed towards some of the effects of climate change resulting from the 'Greenhouse' phenomenon. Scientists have predicted that sea levels could rise significantly in the next 60 year period. The predicted rate of increase is much greater than previously experienced and the rise has been put as high as 1.6m. This high estimate has yet to be confirmed and at an International Workshop in September 1987 at University of East Anglia (5) the report from the group revealed that changes in sea level could be only about 0.6m by 2050.

If the sea level rise approaches that currently predicted it is estimated that some £8bn would have to be expended to reconstruct flood defences and reinstate drainage standards throughout the country.

What is more certain however and has been recognised from recent studies by Draper et al (6) is that there is an increase in the number of storms in the North Atlantic. The intensity of the waves is giving cause for concern and may well have implications for the effectiveness of coastal defences..

Any rise in sea level will most likely result in an increase in depth of water at any structure. As a consequence of this greater depth the wave energy will be increased and either the reflected wave will cause rapid erosion of the foreshore or greater damage will occur to the flood defence structure, particularly if it is an earthen bank or soft cliff face.

At present the allowance in the design of sea defences for the differential rise in sea level is normally 0.3m per 100 years. However some sources consider this to be double a recently assessed rate of 0.15m per 100 years and up to now therefore that allowance has been more than adequate. More complex monitoring programmes will be required to enable researchers to forecast more accurately climatic and sea level changes.

MAFF is liaising closely with other government departments, research institutions and universities to consider strategies to mitigate against these potential hazards. A new standard or suite of allowances for use in designs will need to be derived in the near future.

Materials

In the report of the Waverley Committee reference was made to the need to investigate the use of materials such as asphalt/concrete to waterproof and protect embankments subjected to overtopping. Derivatives of these materials are being employed and proving successful, thanks to the entrepreneurial activities of consultants and contractors.

Great advances have been made since 1954 in the use of polymer materials for reinforcing the vegetation on embankments. During this period, other research has resulted in the selection of grass species which are more resilient and resistant to damage. A report, by the Construction Industry, Research and Informtion, gives guidance on the use of negotiations in civil engineering structures. It must be recognised that all research is expensive but the rewards to be gained can be significant. For instance, forms and shapes of precast concrete units are being developed which remove energy from waves. Those forms range from relatively thin concrete block units, possibly tied to facilitate rapid installation, to deep multi-sided hollow units. Precast concrete units, originally designed as multi-layer units developed for use on structures sited in extremely deep water locations, are being modified for use in the less aggressive coastal regions. If these developments result in lower-cost solutions then clearly they merit further consideration. But to issue cautionary notes, such units should only be used after extensive testing in prototype situations to establish limiting criteria which affect both design and operational expectations.

Much is known about the benefits of using shingle for beach recharge and whilst the most suitable material is won from sea bed locations the material is generally in short supply and can be obtained only from licenced sites, in fierce competition with the needs of the aggregate industry. Availability of material from these sites may

LEGISLATION AND POLICY

be constrained by other interests, especially where they coincide with fish-spawning grounds.

At some stage the material used for beach recharge is passed through pumping equipment and the difficulties and costs of pumping shingle are appreciable. The risk of extensive damage to all parts of a pump is becoming recognised as a major problem which can prove expensive.
It may well be cheaper and easier in the long run to use sandier material, placing excess quantities on the beach to satisfy littoral drift demands. This is an area deserving further consideration.

Unfortunately, re-charged beaches may make frequent demands upon plant and labour to recover and replace displaced material. For the time being, the annual costs remain a maintenance charge and therefore are not eligible for grant aid. However, some reimbursement does occur due to the various allowances made to authorities through the rate support grant system.

Coastal Zone Management

In the departmental report the Waverley Committee commented that much of the damage done by the 1953 flood had been "...the result of sporadic and ill-considered development near the coast which has led to unnecessary expense...by way of additional expenditure on restoration and improvement works".

The Committee commented that the planning machinery had been available to prevent such development and the Committee recommended that "...all possible steps should be taken to prevent further undesirable development..."

Whilst this comment was made in the context of sea defence it must surely be considered in the widest possible context. Similar comments are being made today as a consequence of the need to provide coast defence structures at the public expense to protect developments on unstable areas.
The creation of coastal zones free of development is considered by other authors in following papers.

Regional Coast Committees

The Coast Protection Act provides for the setting up of Coast Protection Boards but none has been set up based upon this provision. However, following on from a research project, by Hydraulics Research Ltd., into the establishment of independent coastal cells, which MAFF inherited from DOE in 1985, MAFF have encouraged local authorities to create Regional Coastal Committees which involve the relevant water authority and other national institutions. To date 5 committees have been set up with the object of exchanging and pooling technical knowledge and data and developing co-ordinated strategic plans.

The establishment of such committees was also recommended by the Institution of Civil Engineers, Coastal Engineering Research Panel in 1985. This move has been strongly supported by the Institution's Coastal Engineering Research Advisory Committee.

The authors of a later paper outline the benefits of such committees.

Summary

Technical standards laid down by the Waverley Committee and the Flood Protection Research Committee provide adequate guidance for the design of flood defences. The latter committee's recommendation provides for defence standards to be adjusted as demand arises.

Researchers are predicting a rise in sea level greater than the allowances currently included in design. These allowances need to be kept under review to ensure adequate protection wherever they can be justified.

Coast defence structures have to withstand similar forces and the criteria used in their design will have to be kept under review.

Legislation affords increasing protection to the environment. Engineering solutions should not be unsympathetic with the geological forms and frequently solutions may provide some enhancement to the natural environment and amenities of the area.

Opportunities exist to use new materials or use existing material in new ways. Provided the resulting schemes are technically sound, cost-effective and the solutions relevant to the needs then such schemes will be considered for grant aid.

MAFF fund extensive research programmes into coastal and marine topics. The results are drawn upon in wider and far reaching research programmes. The Ministry's programmes provide background data to assist with answers to the questions surrounding the effects of climate change and will enable the formulation of national policies needed to determine new policies for flood defence standards into the next century.

LEGISLATION AND POLICY

References

1. Report of the Departmental Committee on Coastal Flooding. HMSO 1954

2. Flood Protection Reseach Committee, Quinquennial Report 1974 to 1979. MAFF 1980

3. Town and Country Planning (Assessment of Environmental Effects) Regulations 1988. SI No 1199. HMSO

4. The Land Drainage Improvement Works (Assessment of Environmental Effects) Regulation 1988. SI 1217. HMSO

5. Climatic Research Unit, University of East Anglia, International Workshop on Climatic Change. Sea Level, Severe Tropical Storms and Associated Impacts, 1987.

6. DJT Carter and L Draper, Nature, Vol 332, 494, April 1988.

4. A county council's approach

P. BELL, BA, MRTPI, Assistant County Planning Officer, Hampshire County Council

SYNOPSIS. This paper describes Hampshire County Council's approach to coastal planning and management, and highlights some current issues, which are likely to have implications for future policy.

INTRODUCTION

1. It so happened that I wrote the bulk of this paper whilst spending a few days on the Cornish coast. As I gazed at the magnificent landscape of the Lizard Peninsular, I thought how clear the objectives for planning in that area seemed to be compared with the Hampshire situation. Here quite obviously was an outstanding national landscape and tourist resource, a primary role which was universally agreed and reflected in its designation as a Heritage Coast. I am not trying to suggest that coastal planning in Cornwall is easy. Indeed, I know personally that proposed development there can be as controversial as anywhere else. The point I am making is that coastal planning and land management can proceed in the context of agreement at national and local level that the conservation of the landscape character is of paramount importance.

2. We in Hampshire feel that our coast is also part of the national heritage and an important recreation resource, and that similar priority should be given to the conservation of its character. Our problem in translating this sentiment into policies which stick stems from the sheer complexity of the area. Within a straight line distance of only 35 miles, the mainland coast contains, in addition to two Areas of Outstanding Natural Beauty and Sites of Special Scientific Interest, two cities of 200,000 people, two major ports, a naval base, the largest oil refinery in the country and 17 marinas. About a million people live within the coastal belt. With a rapidly growing local economy, the scope for conflict is enormous. Against this background, how can conservation objectives be realistically pursued?

COASTAL PLANNING

3. An essential initial step is to be clear about what constitutes the character which is to be conserved. This may sound obvious, but very often planners have neglected to carry out this task. Consequently, many planning documents fail to convey adequately a feeling for the character of the area they cover, and their objectives and policies are bland and platitidinous. This can be vitally important at a public inquiry, where the inspector may be visiting the area for the first time.

4. The key to understanding the character of any area is to identify the things that make it special. To my mind the feature which distinguishes Hampshire's coast is the combination of a very pleasant environment for recreation with a visual scene of exceptional variety and interest. The main elements in this tapestry are:

(a) A varied natural landscape. The New Forest, the many creeks and harbours, and the backdrop of the Isle of Wight give the landscape a special quality.

(b) Large sea-bird populations. The Solent supports large numbers of sea-birds and is of international importance in this respect. It is likely to be designated a Ramsar site.

(c) Historical associations. The Solent has featured prominently in the nation's history, especially with regard to the defence of the realm. This has left a rich legacy of tradition and buildings. The fortifications and naval establishments of different periods are particularly important features.

(d) Ships and boats. Commercial, naval and recreational craft are an essential part of the scene and their movement provides constant visual interest.

(e) Built-up areas. Parts of the built-up coastline, including commercial and naval areas, contribute significantly to its character.

5. The other outstanding feature which singles out Hampshire's Coast from other areas is the opportunities it offers for sailing. The combination of relatively sheltered deep water, a large number of harbours with different characteristics and a pleasant environment make the Solent unique as an area for sailing and other water sports. Its importance as a national and international recreation area is emphasised by its accessibility to the large populations of London and the Midlands.

PLANNING POLICIES

6. A number of considerations have been important in the evolution of planning policies for the coast. These are:

(a) The factors, which make up the character of the coast, are held in a fragile balance which could very easily

be irretrievably upset by thoughtless planning or a single dominating development.
(b) The open coastline has been lost to development at a rapid rate in the last 150 years. About 15 per cent of the mudflats and saltmarshes has been reclaimed in that period. About half the coastline is now built-up.
(c) The pressure for further development, particularly for housing and recreation, is as intense as ever.
(d) The flatness of the coastal plain means that the hinterland, particularly the river valleys, is often important to the character of the coast.

7. Against this background the County Council gives the highest priority to preventing development on the open coastline and on areas clearly visible from the coast, such as Portsdown Hill. The main coastal policy in the South Hampshire Structure Plan states quite baldly that no development should be permitted which would prejudice the landscape or nature conservation character.

8. This is not always such a popular stance as might be imagined. It is not uncommon for the County to take a different view from the District Council on a proposal for a minor recreation development, or a few houses on an untidy, unused corner of a boatyard. To oppose such proposals can sometimes appear unreasonable, spoil-sport, or elitist. Yet bitter experience has shown how easily a few minor proposals can add up to a major change in the character of the coast. The County Council is able to provide District Councils, which are faced with determining applications for development on the coast, with a wider perspective to balance local pressures.

9. Planning policies, which were completely restrictive of further development, particularly recreational development, would be unlikely to succeed completely. Indeed, the pressure for some waterside development stems from an identified need, rather than sheer market demand. Fortunately, there is scope for redevelopment within the existing built-up area, for example, at the Camber, Old Portsmouth, and the Eastern Docks in Southampton. Provided they contain the right mix of uses, redevelopment proposals for these areas will enhance the character of the waterfront and open it up for public access. It is extraordinary how many coastal towns turned their backs on their waterfronts during the first half of this century. The reversal of this trend, which is taking place in many areas, is most welcome.

10. The Structure Plan places one significant restriction on the redevelopment of existing built-up waterfronts. To avoid unnecessary pressure for further development on the open coastline in the future, there is a policy which aims to restrict development on waterside sites to uses, which require a waterside location. This policy can present the Planning Authority with a difficult dilemma. The demand for such uses may not exist at the time that a redevelopment scheme is being considered, or, if it does, it may still not be possible to

COASTAL PLANNING

put together a viable proposal. On the other hand, if waterside sites are taken up by uses which do not require such a location, it may be more difficult at some future date to resist pressure to develop the open coastline.

11. The Structure Plan's policies for the coast are supplemented by a range of policy documents. A statutory Subject Plan has been prepared for the River Hamble by the County Council, which is also the Harbour Authority. Strict policies were required to limit the number of moorings and the spread of development, and the plan, which was the subject of a public inquiry, has proved to be an effective tool in this respect. Non-statutory plans for some of the other harbours, e.g. Portsmouth Harbour, have been prepared jointly with the district councils concerned. The County Council has also published several documents, amplifying its policies, and is about to prepare a coastal strategy, which will take this approach a stage further.

12. The limitations of the powers of the planning authorities are being increasingly exposed. The County Council has already made representations to central government that its jurisdiction should extend beyond the low water mark. The powers conferred on Harbours and Water Authorities are also relevant in this respect.

Fig. 1. The naval heritage is a key element.

THE COASTAL CONSERVATION PANEL

13. Underlying Hampshire's approach to the coast is the realisation that policies on their own are not enough. If

planning objectives are to be achieved, the policies need to be backed up by a variety of measures on the ground, many of which require the County Council to take an initiating, rather than a purely reactive, role.

14. A result of this philosophy has been the setting up of a special panel as a sub-committee of the Policy and Resources Committee to deal with coastal issues. It is chaired by the Leader of the Council and consists of leading members. At least half the members of the Panel are also on the Planning and Transportation Committee. The main thrusts of the Panel's work are:

(a) to work with District Councils, other authorities and landowners to ensure as far as possible that strategic planning policies are consistently interpreted, that opportunities for improving the coastal environment are taken and that coastal land is managed in an appropriate way.
(b) to purchase land and manage it, giving conservation priority.
(c) to maintain a high profile for coastal issues.

COORDINATION

15. A feature of the Solent area is the large number of authorities involved with the coast. There are three County Councils, nine District Councils, Associated British Ports, the Ministry of Defence, ten other Harbour Authorities and two Water Authorities. The need for coordination between these authorities and with other organisations, such as the Nature Conservancy Council, is obvious, yet no authority is charged with this role. The strategic planning responsibility of the County Councils is, I suppose, the nearest approach.

16. The need for a coordinated, comprehensive approach to coastal planning and management is being increasingly acknowledged in many quarters. If County Councils survive, they will undoubtedly have a key role in this respect. It is a role which Hampshire has already shown itself willing to play, for example, by organising two Solent Sailing Conferences.

LAND OWNERSHIP

17. The pattern of land ownership has been a very significant factor in determining the character of the shoreline. For example, most of the land between Lymington and Calshot is owned by a few large estates. This has ensured that the conservation and enhancement of the landscape has been a consistent aim. It has also resulted in very restricted public access in many places. In the Portsmouth area, the Navy's shore establishments are often key sites, from several points of view. They may contain listed buildings and attractive, open landscape in otherwise intensively built-up areas, and in many cases are of considerable historic interest.

COASTAL PLANNING

18. The proper management of land is absolutely crucial if the character of the coast is to be conserved. The County Council encourages landowners to draw up and agree management plans, and may provide assistance in the form of either advice or grant aid. In some cases, it may have a more detailed involvement, for example, by carrying out environmental improvement schemes. The management plans provide a basis for sharpening up conservation policies and, where appropriate, for improving public access.

19. The advantages of owning land, in order to control its use and management, were recognised by Hampshire County Council many years ago. In the 1950s, following a public outcry about a proposal to build an oil refinery, it purchased a large part of the shore between Warsash and Hill Head in order that it should be protected from further development proposals. Many other areas, ranging from Country Parks at Lepe, Netley and Upper Hamble, to nature reserves, to open farmland, have been purchased subsequently. The importance of recreation use is determined by the characteristics of the site, but in all cases conservation is a prime objective.

20. The County Council now has a policy to consider buying any open sites on the coast which come onto the market. It now owns over 3,500 acres of coastal land, 1,300 acres of which has been purchased since 1974.

21. Nearly all the County Council's coastal land holding is now managed by its Recreation Department, which is continually having to adjust as new areas come under its control. The need for close links between the Planning and Recreation Departments is readily apparent.

HISTORIC SITES AND BUILDINGS

22. I have already indicated the importance of the many historic sites and buildings to the character of the coast. The Solent forts and those surrounding Portsmouth and Gosport are obvious examples, as are many of the Navy's shore establishments. A number of sites have come onto the market over the years and further sites are expected to follow in the near future.

23. The County Council has played a key role in trying to ensure that the character of these sites is conserved to the benefit of the environment generally. For example, the Council recently purchased Fort Nelson and Fort Gilkicker and is restoring them. The former now houses the Tower of London's collection of historic armaments. The Council also was actively involved with the Ministry of Defence in initiating a study of land held by the armed services with a view to identifying conservation issues.

24. Ongoing liaison at a high level with the Ministry of Defence, which is continuously reassessing the need for its land holdings, is essential and is given a high priority. Important sites are regularly made available and the planning authorities must decide, and make known, their requirements at an early stage. Like other government departments, the

Ministry of Defence is under instructions to maximise the capital receipts from land disposals.

25. Frequently, the County Council's initial task is to persuade local people of a site's potential for enhancing the character of the area. Often the sites have previously been completely inaccessible to the general public, or the historic buildings and features have been obscured by a rash of structures hurriedly erected during wartime years. After the last war the ancient fortifications on the shore at Old Portsmouth were covered by rows of huts. It required a vigorous campaign by local enthusiasts to persuade the powers that be to restore the historic features, rather than to redevelop the whole area.

26. A current example is the 100 acre site at Priddy's Hard, Gosport, which is about to be disposed of by the Ministry of Defence. Most Gosport people will never have seen the site, yet with some imaginative planning it has the potential to become an attractive and fascinating focal point for the town and the coastal scene. Another example in Gosport is the untidy, little-used area at Haslar, which spoils the entrances to the submarine museum and the imposing hospital buildings.

CURRENT ISSUES

27. Finally, there are a number of current issues, which raise questions about future coastal policies and land management.

The Fawley B Power Station Proposal

28. The proposal for a massive coal fired power station and jetty at Fawley, which the Central Electricity Generating Board may yet seek to progress, has dramatically highlighted the issue which is central to coastal planning in Hampshire - namely, the balance which should be drawn between conservation, recreation and commercial interests. The objectors, which, of course, include the County and a number of District Councils, have no doubt that the proposed development would so dominate a wide area that the present balance would be seriously upset, thereby gravely damaging the character of the area. They contend that considerations of national need should take into account not only the need for electricity, but also the Solent's role as a national recreation and amenity resource.

29. The proposal has provoked much discussion about the status of the Solent in a national context. Whether the proposal is proceeded with or not, I think it likely that there will be an attempt in due course to achieve some sort of recognition at a national level of the special character of the area.

Coastal Erosion

30. Many parts of the Hampshire coast suffer severe erosion problems, which in many cases appear to be getting worse. For

COASTAL PLANNING

example, in the past any loss of material at Hurst Spit has normally been replaced by fresh supplies naturally transported from west to east around Christchurch Bay. Today, expensive artificial replenishment of the Spit is normally required after the winter storms. The predicted rise in the sea level can only add to the problem.

Fig. 2. Coast Protection works are sometimes piecemeal and ugly.

31. Nobody, who has any contact with coast protection and sea defence problems, can fail to be impressed by the sheer complexity of the subject. The number of variables is bewildering and some, such as the die back of the Spertina grasses, are outside man's control. However, there appears to be a growing acceptance that some of our current problems have been exacerbated by ill-conceived schemes on adjacent stretches of coast.

32. I suggest that one of the main reasons for our current difficulties is the inadequacy of the administrative structure for dealing with these matters. The division of responsibility between the Water Authorities, District Councils and landowners is hopelessly blurred. Consequently, there is no proper machinery for coordination, or for carrying out much needed research. As a layman, I also wonder whether the situation has held back the development of expertise in the subject. The government has missed a golden opportunity to improve matters as part of the reorganisation of the water authorities.

33. Although the County Council has no specific responsibility, it is very concerned at the present state of affairs. It thinks little at having to fund a substantial proportion of the cost of schemes over which it has no influence, especially when there is more than a suspicion that some of them are doing more harm than good. Furthermore, it considers that some schemes are ugly and mar the character of the shore and that others are ecologically damaging.

34. The County Council with the agreement of the District Councils and the Water Authority has taken an initiative by commissioning Dutch consultants to advise on a more coordinated and environmentally sensitive approach. That study is still proceeding. The Isle of Wight County Council has also taken a lead by bringing together the District Councils and adjoining County Councils to set up a standing conference to achieve the same objective. Therefore, notwithstanding central government's failure to grasp the nettle, the local authorities themselves are beginning to take steps towards a sensible strategy.

35. I suspect that there needs to be many years of research both to understand fully the processes at work and to discover whether there are new solutions which would be cheaper and more sympathetic to the environment than some of today's schemes. What should be the balance between major capital schemes and less dramatic solutions requiring on-going revenue expenditure? Could money be saved by more regular maintenance? Are there biological solutions to some problems, for instance, by introducing new strains of marsh grasses which are not prone to die back?

Yacht Moorings

Table 1. Yacht Berths in the Solent

	1973	1978	1985	1988 (estimate)
Hampshire	17703	21640	25764	27126
Isle of Wight	3596	3769	4252	4300
Solent Total	21299	25409	30014	31426

36. Table 1 illustrates the dramatic rise in the number of yacht berths in the Solent area in the last fifteen years. Planning permission has been granted for a further 2,000 and proposals for a number of other marina developments are being actively pursued.

37. Most of the recent developments have included a package of uses, of which the marina, although the focal point, is only a part. Waterside housing has proved to be especially popular. Such schemes form the main part of the redevelopment

COASTAL PLANNING

Fig. 3. Marinas - Recreation haven or developers' goldmine?

of Southampton Eastern Docks, for example, and they appear to be very successful. The planning authorities are extremely wary, however, of proposals by existing marinas and boatyards to introduce alternative uses on part of their sites. For instance, proposals for housing on the River Hamble have been firmly resisted on the grounds that they would mar the character of the river.

38. The continuing growth of the number of yachts is increasingly raising the question of the capacity of the Solent in this respect from several points of view - navigation, the environment and the conflict with other uses, including other types of recreation.

39. A number of individual harbours have placed a ceiling on the number of berths, but there is also concern about safety at certain key points on the open water. Restrictions have already been introduced at the mouth of Southampton Water off Calshot and I suspect that they may have to be considered for other areas before long. Traditionally, yachtsmen are not very amenable to being regulated in this way. There is also the question of who pays for the supervision which will be required.

40. Navigational capacity is, however, only one element in this issue. Environmentalists have been concerned for some time, and now, significantly, even organisations, such as the Regional Committee of the Royal Yachting Association, are questioning the wisdom of allowing further moorings, on the grounds that excessive congestion at the harbours most popular with visiting yachtsmen is seriously diminishing the pleasure from sailing in the area.

41. There is a long way to go before a consensus view emerges. There are many who do not see a particular problem, or who are primarily concerned with the advantages of yachting to the local economy, or who simply feel that the Solent

should be enjoyed by as many people as possible. The next few years are likely to see an interesting debate.

Wardening

42. The coast, probably more than any other area, is under pressure from a growing population with increasing mobility and leisure time. Many of the problems do not arise from a lack of long term planning. They simply require an immediate presence on the ground. Both County and District Councils are already involved in wardening particular stretches of heavily used coastline and I am certain that this service will have to be extended, if the situation is not to deteriorate.

43. The cost of this will be a problem and therefore the positive side of a wardening service needs to be constantly emphasised. For example, the ability to identify, and deal with, maintenance problems at an early stage can save money in the longer term. The prospect of clearing up some of the disgusting litter on our beaches is also surely a good selling point.

44. A short paper can only give an outline of the County Council's approach to coastal planning and management. We do not claim to have all the answers by any means, but I do believe that our experience may be helpful to others in different parts of the country. Above all, I hope that I have been able to convey the commitment and enthusiasm of both members and officers in Hampshire to coastal conservation. Without that, there is no hope of success.

5. Physical constraints on the planning of coastal development

D. BROOK, BSc, PhD, FGS, Head of Land Stability Branch, Minerals Division, Department of the Environment

SYNOPSIS. The physical condition of land, including such features as landslides, subsidence, erosion and flooding, may impose severe constraints on the development and use of that land. Planning guidance on 'Development on unstable land' advises planners and developers on the consideration of such constraints in their decision-making. In so far as they affect land use, physical constraints on land may be material planning considerations. The responsibility for determining whether land is suitable for a particular purpose remains, however, with the landowner/developer. This general guidance is here examined with specific reference to the coastal situation.

INTRODUCTION

1. The purpose of the planning system is to regulate the development and use of land in the public interest. The fundamental requirement of the legislation is that development may not be undertaken without planning permission.(ref.1)

2. In deciding whether to grant permission, the local planning authority must 'have regard to the provisions of the development plan, so far as material to the application, and to any other material considerations' (Town and Country Planning Act, 1971). Other material considerations are not specifically defined and can cover a wide field. However, the system is concerned with land-use planning matters, ie those relating directly to the physical development and use of land and not to other matters.

3. There are many constraints on the development and use of land which need to be considered by planners and developers. The nature and form of the development, the location of a site and its physical, social and environmental characteristics may all impose constraints.

4. Planners are accustomed to considering such features as landscape conservation, green belt, access limitations and the impact of traffic, and the environmental effects of noise, dust and waste disposal. They may be less accustomed to considering physical constraints on land such as landslides, subsidence, erosion and flooding.

5. Risks to life and health, damage to buildings and struct-

COASTAL PLANNING

ures, and loss of industrial production may all arise if inappropriate development does not take due account of inherent physical constraints. The result is often considerable public alarm and requests for Government funding of remedial or preventative measures and/or compensation for losses suffered.

6. Experience has shown that the planning system ought, in the public interest, to take more account of physical constraints on land. Planning guidance on development of unstable land has been prepared by the Department of the Environment. This advises on the circumstances in which consideration of physical constraints might be needed and how it could be applied in practice. More specific guidance may be needed in the coastal situation and this is currently under consideration. This paper is intended to examine some of the issues that might arise.(ref. 2)

PHYSICAL CONSTRAINTS ON LAND

7. Britain has a wide range of physical conditions of both natural and man-made origin which may impose constraints on the development of land. Four broad categories can be defined:-
   (a) underground cavities; these may be natural cavities, such as those formed by solution, tension gulls due to cambering or sea caves, or result from mining or civil engineering works;
   (b) unstable slopes; these may be natural slopes or man-made, whether excavated as in quarries and cuttings or constructed as in tips and embankments; instability may result from natural processes and from man's activities;
   (c) ground compression; both natural ground, underlain by peat or soft alluvial, estuarine or marine soils, and made ground, such as landfill or backfilled quarries and tips, may suffer excessive compression; shrinking and swelling of clays and extraction of ground fluids may also cause significant ground movement;
   (d) flooding; both rivers and the sea may cause flooding, leading to erosion, deposition of sediment and general disruption; man's activities may increase or decrease the impact.

8. The geographical distribution of such constraints and their effect on surface land use are variable and may be imperfectly known. All four categories of constraint may be present in the coastal situation. However, the principal concern must be with slope instability and flooding in this most active geomorphological environment. For discussion purposes, further comment is largely restricted to unstable slopes and coastal erosion but the principles generally apply to all types of physical constraint.

Coastal erosion and landslides

9. The United Kingdom has about 17000 km of coastline of varying types. Some is eroding, at rates up to over 3m per year, some is essentially stable and some is accreting. A

current EEC project is looking at morphosedimentological segments of coast and the dynamics of thoe segments with the aim of bringing together available information on the present state of the European coastline. It will indicate the range of cliff types and erosion situations in both hard and soft rocks and will provide a useful first source of information for planners. (ref. 3)

10. The Department of the Environment review of landsliding in Great Britain identified 850 landslides along the coast. Of these, some 582 were in 'rock', 155 in 'superficial materials' and 113 involved both 'rock and superficial materials'. However, this review considered only recorded landslides and did not attempt to identify landslides not recorded in the published literature. There is, therefore, a considerable under-representation of landslides. For example, the landslides review recorded one landslide along the coastline of Torbay. During a recent ground survey, some 304 landslides were identified and mapped. (refs. 4,5)

11. In general, there is a danger of landsliding, in one form or another, along every cliffed portion of coastline in the country. Analysis of the published records, however, indicates that some strata are more vulnerable than others to coastal landsliding. Most vulnerable would appear to be coastal outcrops of glacial deposits and London Clay; more vulnerable than most are outcrops of Lower Lias and Upper Greensand/Gault; and more vulnerable than many are outcrops of Chalk and Upper Carboniferous strata.

12. Most of the coastal landslides recorded in the literature are currently or recently active, ie with movement having been recorded in the last 100 years, and the predominant cause of coastal landsliding is erosion. Change in water regime is also an important, though not dominant cause. It is important to note, however, that there is a good deal of erosion taking place which is as yet undetected at the cliff top in many areas. Such cliff top lag may be very significant and cliff top erosion may continue for many years after toe erosion has ceased. For example, cliff erosion is continuing at Shanklin, Isle of Wight, over 100 years after development of the Esplanade at the foot of the cliff. (refs. 6,7)

PHYSICAL CONSTRAINTS IN PLANNING

13. The responsibility for determining whether land is suitable for a particular purpose rests primarily with the developer. It is for the developer to investigate and assess the stability of a site to ensure that it is stable or that any instability can be overcome by appropriate remedial, preventative or precautionary measures. It is not the function of the local authority to investigate ground conditions unless they propose to develop a site themselves.

14. The local planning authority is, however, empowered to control most forms of development under the Town and Country Planning Acts. In so doing it has a duty to take all material considerations into account. In so far as it affects land use,

the stability of the ground may be a material consideration which should be taken into account when deciding a planning application. The role of planning control is to determine whether a development should proceed; having decided that it should, the role of building control is to determine how it can proceed safely.

15. The planning system acts in the public interest, not to protect the private interests of one person against the activities of another, or even against the consequences of his own activities. However, it is clearly in the public interest that unsuitable land should be avoided where possible. Planners should , therefore, be aware of the potential physical constraints on development in their area and should take account of them in policy formulation and in decisions on individual applications.

Development Plans

16. Development Plans provide the broad framework or rational and consistent decisions on planning applications. Structure Plans set out the strategic policies and proposals and Local Plans give detailed expression to those policies and general proposals. In preparing and altering their Development Plans, local planning authorities need to take account of the physical constraints of which they could reasonably be expected to be aware.

17. Any survey preparatory to drawing up Development Plans should identify as far as possible the physical constraints on land within the plan area. Areas subject to flooding, coastal erosion and landsliding should be identified and the plan policies and proposals should take account of these factors. The EEC project on coastal erosion and the National Landslides Database provide a minimum of information but, often, much more information will be available locally. In some circumstances, consideration will need to be given to the commissioning of specific work to identify the constraints and to assist in dealing with them.

18. The Department of the Environment is currently engaged in research on the Isle of Wight undercliff at Ventnor to develop a method of landslip potential assessment which is generally applicable to areas of coastal lansliding in successions with interbedded poorly lithified and stronger rocks. Applied geological mapping has also been funded by the Department in the Torbay area and elsewhere. Hazard zonation of coastal landslides has been undertaken by the University of Exeter for the District Councils of North Devon, South Hams and East Devon.These and similar techniques merit consideration for use by local authorities. (ref. 8)

19. Having thus identified the nature and extent of potential problems within their area, the local planning authority should take due account of them in policy formulation. The aim is not that physical constraints should over-ride all other considerations but that they should be considered alongside them. The

Structure Plan might include a general policy statement that physical constraints on land will be treated as material considerations in determining planning applications. The Local Plan should then set out in detail the criteria which will be used in determining planning applications and the planning conditions normally expected to be met. Development proposals should also take account of physical constraints.

20. The planning system has an in-built presumption in favour of development, having regard to all material considerations, unless that development would cause demonstrable harm to interests of acknowledged importance. In areas of active coastal erosion, landsliding or at risk from flooding, the public interest may dictate a reversal of this presumption. In greenfield areas the avoidance strategy represented by a presumption against development may present little difficulty. It will often in such cases be supported by the conservation criteria considered in DOE Circular 12/72 (The Planning of the Undeveloped Coast). (ref. 9)

21. In areas of existing development, a presumption against new development because of the threat from coastal erosion and instability or flooding may also be justified. Careful consideration will be needed, however, because of the inevitable conflict between the interests of existing property owners and the public interest in terms of rational planning decisions. One consequence could be a depreciation in the market value of existing properties because such a policy might indicate a conscious decision that such properties were not going to be protected. The alternative would appear to be that development would continue until such time as protection works became economically justified, which could have significant implications for the public purse. It is of course the risk from natural processes which would blight property values in this situation rather than the policy itself.

Development Control

22. Save for certain categories of permitted development defined in the Town and Country Planning General Development Order 1988, specific planning permission is required for development upon application to the local planning authority. The decision whether to grant planning permission will need to take account of the physical constraints alongside other planning considerations.

23. Where the local planning authority has good reason to believe that coastal erosion, instability or flooding could make the ground unsuitable for the proposed development or could adversely affect it or neighbouring land as a result of the development, it will wish to be satisfied that the developer has taken adequate account of such factors in his proposal. Only then can the authority make a sensible decision.

24. Even before an application is made, informal discussions between a potential developer and the local planning authority can be very helpful. Any constraints on the proposed development can be brought to the developer's attention at this stage

and the implications explained. The applicant can then design his scheme so as to take full account of the likely requirements of the planning authority.

25. If the information initially provided by the applicant is insufficient for proper consideration of the physical constraints on a proposed development, the local planning authority may take the same steps as it would with regard to any other planning consideration. It is empowered under the General Development Order to require the developer to provide, at his own expense, additional information in the form of a specialist investigation and expert assessment of the physical condition of the site and confirmation that the proposal incorporates any measures recommended as a result of that assessment.

26. The criteria for assessing such information will have been set out in the Local Plan. Provided that the assessment has been carried out by someone with the required expertise and includes all the necessary elements, the planning authority may opt to accept that assessment at its face value. The identification of the required expertise and of the necessary elements may require further guidance. Alternatively, if it has or has access to the required expertise, eg from other parts of the local authority or by employing consultants, the planning authority may prefer to assess in more detail such accompanying information before determining the application.

27. In some cases, the developer's specialist investigation and assessment may demonstrate that the land is stable or that the level of instability is such that it will not adversely affect the proposed development or neighbouring land. Determination of the application may then proceed solely on other planning considerations.

28. In other cases, the developer's investigation may indicate that certain measures need to be taken in order for the development to proceed. In such circumstances, the application should include such measures in the design of the development. Any permission granted should be conditional on the incorporation of these measures in the development. It is also open to the planning authority to phase the development by the use of conditions so that any treatment should be carried out before the activity, against the effects of which it is designed to protect, shall take place.

29. There will also be cases where the instability is such that it cannot satisfactorily be overcome and the local planning authority may have no alternative but to refuse permission. For example, the rate of coastal erosion may be such that the development would be threatened in the short or medium term and could not therefore have a reasonable life expectancy or the risk of flooding may be such that it would not be possible to secure the safety of occupants. Alternatively, the necessary remedial measures may have an adverse impact on interests of acknowledged importance, or may be on land which is not in the control of the applicant.

30. Permitting development with a 'sea view' may be

eminently desirable as far as the developer is concerned but if that development is on eroding sea cliffs then it may have significant implications for the public purse in demands for coast protection works. Local planning authorities need to be aware of these implications and of potential consequential effects elsewhere along the coastline in determining such applications.

31. Occasionally, the information supplied by the applicant will be sufficient to determine whether the development should proceed but insufficient to resolve specific details. It may then be appropriate to grant planning permission subject to conditions that the development will not be permitted to start until an adequate investigation and assessment has been carried out and that the development will need to incorporate all the measures shown in that assessment to be necessary. It is unlikely, however, that such conditions would be appropriate in the case of active coastal erosion or landsliding.

SUMMARY AND CONCLUSIONS

32. Development in coastal areas is subject to a number of physical constraints. Principal among these are coastal erosion landsliding and flooding. Planners should be aware that such constraints may be material planning considerations. They need to be considered at all stages in the planning process. Surveys for forward planning should identify as far as possible the nature and extent of physical and other planning constraints and Development Plan policies and proposals should take due account of them. When considering individual applications, the planning authority should satisfy itself that the necessary consideration has been given to such constraints alongside other planning matters with which it may be more familiar.

33. Within the context of Government policies and the provisions of the Development Plan, each application for planning permission should be decided on its merits. The way that this is done is a matter for the local planning authority to determine but it should do so in full knowledge of the potential consequences of its decisions.

34. DOE planning guidance on 'Development on unstable land' advises on general principles with respect to considerations of all forms of instability in the planning system. More specific guidance on coastal planning and development may be needed and the provision of this is currently under consideration. Particular issues that may need to be addressed may include the need for consideration of physical constraints within the context of the whole active coastsl system and the effects of rising sea levels, possibly due to the 'greenhouse effect'.

35. A number of consequential effects may arise from this necessary consideration of physical constraints in planning, particularly in respect of any presumption against development in areas of actvie coastal erosion and the implications for the public purse of permitting development on eroding sea

cliffs. The effects on property values and issues in relation to permitted development and purchase notices will need to be examined very carefully.

36. Whatever consideration is given to physical constraints within the planning system, it is important to remember that the responsibility and any subsequent liability for safe development and secure occupancy of a site rests with the developer and/or the landowner/occupier. In particular, the granting of planning permission does not give a warranty of support or stability.

ACKNOLEDGEMENTS

37. The preparation of this paper has been assisted by many discussions over the years with colleagues both within and outside the Department of the Environment, to whom thanks are due. The views expressed in this paper are, however, those of the author alone and do not necessarily represent the views of the Department of the Environment or any other organisation.

REFERENCES
1. DEPARTMENT OF THE ENVIRONMENT Planning policy guidance: General policy and principles. PPG1,Jan 1988, HMSO,London.
2. DEPARTMENT OF THE ENVIRONMENT Minerals planning guidance: Development on unstable land. Consultation draft, Aug. 1988, DOE, London.
3. MAY, V.J. Spatial variations in rates of cliff erosion and the response of cliffs to changes in the intensity of erosion of the cliff toe. (Abstract) Geol.Soc.Newsletter, Vol.17, No.6, Nov 1988.
4. GEOMORPHOLOGICAL SERVICES LTD. Review of research into landsliding in Great Britain. Report to Dept Environ. 1986-87.
5. GEOMORPHOLOGICAL SERVICES LTD. Planning and development: Applied earth science background, Torbay. Report to Dept Environ. GSL, Newport Pagnell.
6. BARTON, M.E. & L.D. MOCKETT The abandoned cliffs in the ferruginous sands at Shanklin, Isle of Wight. (Abstract) Geol.Soc.Newsletter, Vol.17, No.6, Nov 1988.
7. MADDRELL, R.J. An examination of abandoned and eroding cliffs: Shakespeare Cliffs, Folkestone and Fairlight Cove. (Abstract) Geol.Soc.Newsletter, Vol.17, No.6, Nov 1988.
8. GR NGER, P. & P.G. KALAUGHER Hazard zonation of coastal landslides. In Landslides, Vol.2, Proc.5th Int.Sympos. on Landslides, 10-15 July,1988. A A Balkema, Rotterdam.
9. DEPARTMENT OF THE ENVIRONMENT The planning of the undeveloped coast. DOE Circular 12/72 (WO Circular 36/72).Feb 1972. HMSO, London.

(c) British Crown Copyright 1989.

This paper is reproduced by kind permission of the Controller of Her Majesty's Stationary Office.

6. A regional strategy

B. HALL, FICE, FIWEM, Consultant

SYNOPSIS. Provision of coastal works is the prerogative of multiple coastal authorities and this is necessarily so. However to be effective each authority needs national and regional data on which to plan and execute its works. The present ad hoc information gathering needs to be replaced by an integrated continuous effort on a national and regional basis. The nature of the coastline is such that extensive lengths require defence for the first time or, more commonly, renewal of existing works. A rare opportunity for change is available. Identification of need and long term planning is necessary if the coastline is to be effectively managed.

NATURE OF THE UNITED KINGDOM COASTLINE

1. The length of the U.K. coastline is not particularly relevant but the value of works needed to protect the coastline from erosion and prevent flooding from the sea is. The sea walls provided are reported by CIRIA as an asset worth £4000m. requiring some £29m. worth of annual maintenance and having a life of fifty to one hundred years (1985 prices).

2. Even at 1985 prices either the figures are low or the present need for renewal is substantially greater than the average. Anglian Water report a need for investment of £150m. in sea defences over the next five years and other conference papers describe the background to this.

3. The nature of the U.K. coastline is very variable. Much is composed of materials from sedimentary deposits and is subject to almost continual change. The beaches of shingle or sand, dunes, saltings, marshes etc are characteristics recognised as representative of a substantial proportion of U.K. coastline. It is these soft interfaces with the sea that are stabilized by intervention and need investment of perhaps £100m. p.a. plus maintenance of some £35m. p.a.

4. It is the regions with substantial lengths of soft coast line that need to develop a strategy allowing forward planning of necessary rehabilitation, upgrading or new works within a clear management framework.

Change in sea level

5. For a variety of reasons sea levels are changing. In the North Sea there is general acceptance of a slow rise in mean sea level and a marginally greater rise in maximum sea level relative to the land. This change in sea level undoubtedly contributes to the recession of beaches and saltings but is not regarded as the only cause.

6. A rise in sea level increases the risk of flooding from the sea. Authorities responsible for defence against the sea are sensitive to the need to increase the level of defences if they are to maintain the standards of protection against flooding. Whilst there is no nationally adopted standard for defence in terms of the frequency or return period of flooding, economic assessment suggests defences to provide against the 1 in 100 year tide are justified for all urban areas and that this may be unacceptable in highly developed zones. Public expectation is increasing. This facet also increases the demand for investment.

Recession of the coastline

7. All coastlines are subject to attack by the action of the sea and those composed of sedimentary materials show measurable change. Currently the extent of areas being eroded by waves and currents is reported as exceeding the areas that are accreting. The apparent recession of soft areas has been recognised for a sufficient period for short term cyclical change to be unlikely as an explanation. Whilst general recession is perhaps better documented on the east coast the trend is perceived throughout the U.K.

8. Where defences have been built their design has generally assumed the retention of some foreshore materials but the loss has often been greater than expected. The increased wave action and/or changed tidal currents consequent upon loss of foreshore have reduced the anticipated life of sea defences.

9. There remains a wish to retain existing lines of defence both against flooding and erosion but in some areas doubt as to the practicality of this is increasing. It may be that acceptance of land loss will become necessary and authorities will be faced with difficult judgements as to where the coastline should be held and where released to flooding or erosion. It is desirable that such decisions should be made in a regional context if the balance is to be the best available. Where a sacrifice of land - to flooding or erosion - is beneficial to a region as a whole then compensation to property owners might be appropiate but this is a difficult concept to address within the framework of permissive powers for defence works.

Provision of coastal works

10. Other conference papers describe how coast protection and flood defence against the sea are provided for by statute in England, Wales and Scotland and are funded partly from the

national exchequer and partly from local authority rates. The
legislation is permissive, ie, the responsible authority is
given powers to provide defences but no specific direction as
to what defences are appropiate. Works proposed are subject
to approval before any national funding is forthcoming but if
no works are proposed then there is no machinery of compulsion.
Fortunately the public demand is sufficient to ensure reasonable action is taken.

11. The authorities responsible for sea defence in England
and Wales are the water authorities as regards flooding and
the maritime district councils for prevention of erosion and
similar arrangements exist in Scotland. The overall situation
is then perceived as national control with local provision.
The water authorities can be regarded as regional in the extent
of their coastline whilst maritime district councils are much
more localised.

THE INTERFACE BETWEEN AUTHORITIES

12. Against a background of a countrywide problem and
regional or local provision for a solution there is a danger
of piecemeal provision of defence against the sea. The causes
of coastline change recognise no administrative boundaries and
in the past the solution to a local problem has sometimes been
a contributary cause to a new problem on an adjacent coastline.
The better understanding of the interaction between sea and
foreshore has much reduced the risk of this happening. However
data exchange between coastal authorities has not progressed
to a sufficient degree to ensure that the understanding of
foreshore forces is fully shared.

13. Regional evaluation of the forces at work followed by
dissemination of information is needed to improve the situation.

14. Clearly the resources used to understand and deal with
a local problem must be in keeping with the size of the problem.
Conservative design of remedial works can be utilised to offset
the inability to fully research with the time and resources
available. Within a short time or distance another local
problem can be dealt with in the same way and the process
repeat itself. With forethought, co-operation and planning
the work necessary to research a number of local problems
would finance itself from project savings and produce long term
benefits to the coast protection provided.

General information needs

15. To develop the relationship between the coastal
authorities and define how the necessary data collection work
should proceed there is a need for areas of activity to be
defined. There is clearly a need for national and regional
authorities to obtain offshore physical information relative
to any length of coastline requiring defence against erosion
or flooding and to understand the mechanisms, similar information is required for a coastline which is accreting. In
simple terms offshore of all soft coastlines and at their

termination full documentation of change and causation is required. Equally physical information is required onshore where there has been, or is likely to be, intervention to reduce change at the coastline. The onshore data required is generally more specific but does include the offshore effects when translated inshore.

Survey and analysis at national and regional level

16. The first need is for water level relative to land and the influence of long term secular change, tidal cyclic variation and random surge or similar effects. The data together with joint probability analysis is generally available from statistics gathered nationally by Proudman Laborty and others and is suitable for forward projection.

17. Wind data as to strength, direction and duration is also generally available from the Meteorological Office. Offshore wave records giving height, direction and form are generally inadequate but may be sufficient to support hindcasting from wind data. Again joint probability with variation in sea level is relevant.

18. Offshore surface and bed currents are needed and data readily available is usually too general to correlate with other factors. Bathymetry and bed material requires greater survey effort if bed movement is to be understood and forecast with confidence. All offshore physical information together with its interpretation is relevant to long stretches of soft coastline.

19. Landward of the low water mark any water level propagation into estuaries etc is relevant. The wave climate at the coastline is, of course, strongly influenced by the sea bed and offshore imput requires modification by refraction models or other techniques to provide onshore data.

20. For understanding the coastal processes, data is required concerning the nature and inter-relationship of wave climate, beach form and sediment transport both littoral and onshore/offshore. Both the long term changes and the cyclic changes are relevant.

21. Given an understanding of the processes at work, evaluating the need for intervention requires appreciation of the interaction with existing structures. The standards of protection and the projected life of coastal structures are relevant and essential to proper planning of future work. Decisions as to future standards are the prerogative of the coastal authorities but choices have to be made in the light of resources available and other demands upon those resources. To do otherwise means a breakdown of the service or continuous emergency works that are inefficient and rarely cost effective.

22. To aid management in making the choices the extent and nature of the land area at risk is relevant and must be assessed. Benefit/cost ratio is a useful approach but crude in that it emphasizes apparent material benefit and fails to recognise quality of the environment provided. More refined appraisal techniques are needed.

MANAGEMENT

23. Effective management of the coastline requires very extensive data and analysis. Basic information concerning offshore parameters are provided for by the Meteorological Office, Proudman Laboratory and other national organisations. Those coast protection organisations administered regionally are well placed to measure or evaluate wave propagation at the coastline, sediment transport etc but the local coast protection authorities are not. Water authorities should collect and provide this intermediate level of information to the local coastal authorities. In that they generally need the information themselves the additional burden would be small.

24. Each coast protection authority will of necessity have to determine the standards to be provided by the use of their permissive powers. Consideration of such standards will commence with concern for the hinterland. From doubts as to the effects of flooding or erosion will be borne an attempt at evaluation firstly of the consequences and then of the risk. In the light of limited resources, management will want to know the price for defence and the appropiate standard in terms of damage avoided. From such debate will be borne the concept of benefit cost ratios in dealing with either erosion or potential flood risk. A concept may be adopted of providing flood defence at an economic level

25. Only the responsible coastal authority can make decisions as to standards of protection to be provided and it will fall to their staff or consultants to assemble and interpret the data necessary for such deliberations. Such an exercise will commence with the land or property at risk and progress to an assessment of existing defences with the inevitable difficulty of evaluating the life of works likely to be subjected to extreme events. Wave climate, sediment transport, water levels etc are all relevant and synthesis of alternate options will require a partially reiterative process.

26. Such work involves the making of judgements at levels varying from fine detail to overall policy. To be effective a hierarchy of responsibility is needed in which each authority specialist or adviser evaluates and pronounces within their area of responsibility, communicating as necessary and contributing to the overall picture without duplication or omission. In this way information derived at national, regional and local level can be brought together, its relevance determined and the parameters used effectively to evaluate methods of achieving objectives set by management.

PLANNING

27. Planning coastal works starts with identification of need. A plan for maintenance, rehabilitation, renewal or change is required to show the total need, a fifteen year plan for provision and a five year programme of works. Without such a forward looking plan there is little hope of the

COASTAL PLANNING

necessary resources being assembled and the preliminary investigations being completed in time for effective use. Also summation of the improvement or deterioration – initially predicted and subsequently measured – will enhance the quality of the planning and allow review or reassessment at intervals.

28. The permissive nature of the powers and the constraints of funding or other scarce resources will influence the programme of works and some form of priority system is required. The complexity of the subject requires a weighting to be given to the more critical aspects of particular works if their priority is to be uniformly assessed. In this respect there are probably three aspects for consideration at a primary level with many other facets for secondary consideration if such refinement is needed. The following are suggested for primary considerations.
   (a) Assessed life of existing works.
   (b) Shortfall relative to adopted standard.
   (c) Benefit/cost or apparent return on investment including allowance for foreseen development in the hinterland.

29. Consistency in the evaluation of the criteria is essential for any priority system to be workable.

30. A system of weighting as suggested can indicate order for programming and hence works. It can do nothing to resolve the conflict where coastal works compete for resources with the demands from other types of public service. I would not presume to make suggestions for resolving this problem.

31. The need for data collection over a long period cannot be stressed to highly as much of the work depends upon analysis of the records. For confidence those records should be of a duration of the same order as structure life. It would not be inappropiate to commence recording as works come into use against the need for information when contemplating replacement.

32. For the evaluation of probability of extreme events records need to be even longer.

FUNDING AND EXECUTION

33. The funding of works for coastal management is partially from the national exchequer by way of grant aid for approved works. Most of the balance after grant aid is raised by the county and district councils through the general rate system. No proposals for fundamental change to funding are known at this time. The system appears to work but suffers from long term uncertainty which inhibits forward planning.

34. The National Rivers Authority is scheduled to take over the powers presently used by the water authorities in 1989 and will continue to operate on a regional basis. The National Rivers Authority will be ideally placed to build up information on the offshore parameters which both they and the district councils require to plan coastal work needed. Ultimately those bodies having power to carry out works must be responsible for the execution of those powers. Given

national and regional support in data, information, and interpretation there is no reason why their management and long term planning should not be effective.

7. Research on tides, surges and waves

G. ALCOCK, BSc, MSc, BA, MInstP, CPhys, Principal Scientific Officer, Proudman Oceanographic Laboratory, Bidston Observatory

SYNOPSIS. This paper gives an overview of recent and current U.K. research applicable to coastal management. The Institution of Civil Engineers' Report on Coastal Engineering Research (ref. 1) is taken as a starting point to define "recent".

DATA COLLECTION
Sea level data
1. Fig. 1 shows the U.K. national tide gauge network which, since its inception in 1953 after the storm surge, has undergone change both in the sites which constitute the network and the instrumentation installed at the sites. The underlined sites indicate DATARING installations which have a microprocessor-controlled data logger linked by telephone line to the central station at the Proudman Oceanographic Laboratory (POL) at Bidston Observatory, Merseyside (ref. 2). Records are processed at the POL to provide hourly series of observed and surge (observed - predicted tide) data (ref. 3).

2. Further planned work on the DATARING system includes the development of faster data transfer, development of an interface system to allow other users such as the Water Authorities to have access to the system, evaluation of acoustic sea-level instruments and development of inputs from offshore pressure recorders.

3. A number of the east coast sites are equipped with telemetry systems enabling the Storm Tide Warning Service (STWS), based at the Meteorological Office at Bracknell, to monitor sea levels in the North Sea and so give early warning of danger levels. The Aga equipment which provides this information is now being replaced by new equipment designed by the POL.

4. The national network is supplemented in some areas by localised network e.g. Anglian Water Authority's telemetry gauges at Aldeburgh, Great Yarmouth and Wells (ref. 4); Yorkshire Water Authority's telemetry gauges for the River Hull Tidal Surge Barrier (ref. 5); and the Port of London Authority's POLATIDE system on the River Thames (ref. 6).

RESEARCH

Fig. 1. Tide gauges of the national network, December 1988. (Modernised installations are underlined).

Other permanent gauges are operated by other organizations, mainly water or port authorities.

Current data

5. There are no permanent current meter stations operating off the U.K. coast; such data collection is usually carried out during specific coastal studies e.g. for the Wirral off-shore breakwaters (ref. 7). The British Oceanographic Data Centre (BODC) at the POL maintains a databank which includes moored current meter data.

6. Coastal studies usually use conventional rotor, electromagnetic or acoustic current meters; but all these instruments have problems in measuring the surface currents predominantly responsible for controlling the movement of coastally-discharged buoyant contaminants. These currents can be measured using float-tracking, or by recording high frequency radio-waves back-scattered from the sea surface using the Ocean Surface Current Radar (OSCR) recently developed at the Rutherford Appleton Laboratory.

7. OSCR units have been deployed from over 20 sites to provide information up to 40km offshore for engineering design of off-shore works (ref. 8 and Fig. 2). The system can operate under a wide variety of conditions and can measure the amplitude and direction of the predominant tidal currents with an accuracy of 5cm s^{-1} and 5° respectively from as little as seven to ten days of half-hourly recordings. (These values can be significantly improved by more frequent observations over longer sampling intervals extending for a month or more.) Useful information is also obtained on both the net residual surface drift and the wind-driven components. Comparisons have been made of currents measured by OSCR, current meters and drogued floats within one metre of the sea surface (ref. 9).

8. Research and development is underway by Marex Ltd. to produce an OSCR system with a much smaller "bin" size than the existing 1.2km, and to provide miniaturised automated hardware that should make the operational logistics easier.

Wave data

9. Although the I.C.E. Report (ref. 1) recommended a national plan for wave data acquisiton as an overall priority, the routine collection of instrumentally-recorded wave data around the U.K. coast actually has been reduced over recent years to just a few Light Vessels ("Channel", "Sevenstones") or coastal sites (e.g. POL's installation at West Bexington for Wessex Water). However, Anglian Water Authority propose to install a pressure transducer at Boygrift, Lincolnshire to routinely measure wave heights and sea levels.

10. Other measurements are taken for specific studies e.g. recently by the POL and the Institute of Oceanographic Sciences Deacon Laboratory (IOSDL) off Flamborough Head, Holdness and Kinnairds Head (ref. 10), and for wave set-up

RESEARCH

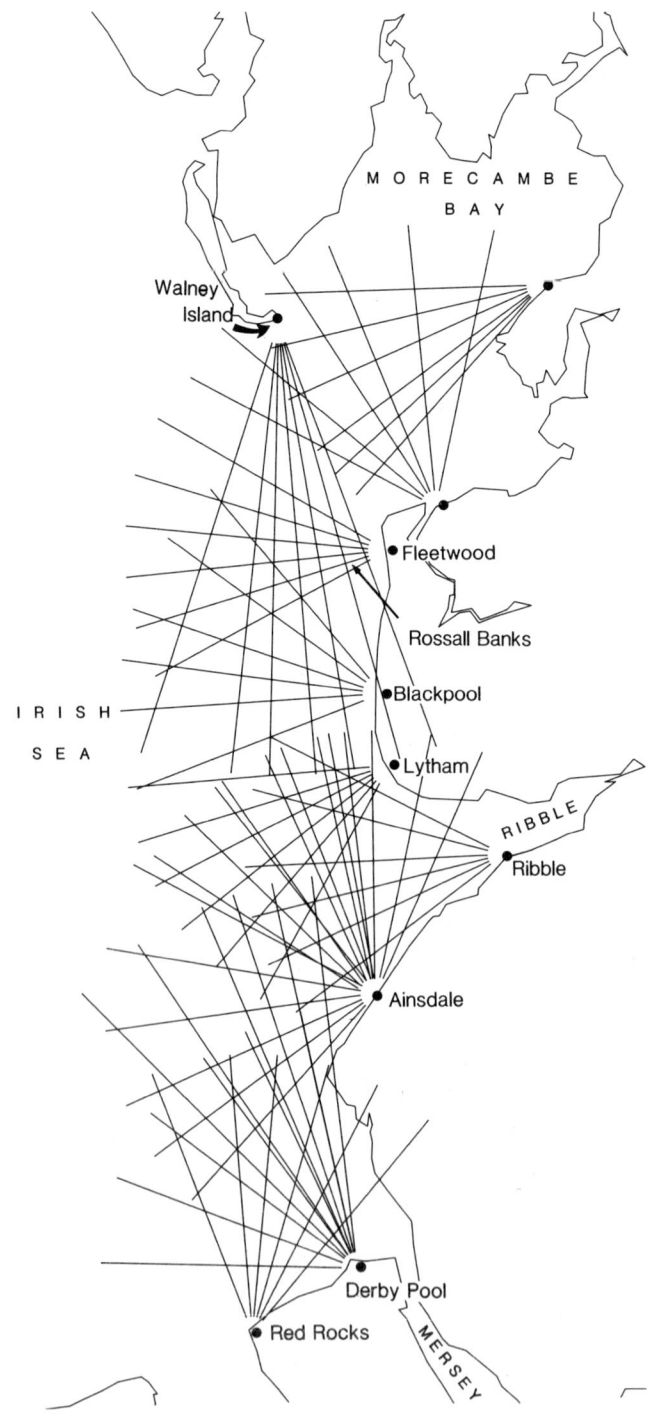

Fig. 2. OSCR coverage on Cumbrian coast.

studies at Chesil, Hornsea and Woolacombe. An inventory of available instrumentally-recorded wave data is published and periodically updated by the BODC.

11. High frequency radar to infer the wave spectrum over a wide area (out to 100km or more) and a satelliteborne synthetic aperture radar (SAR) (for a directional spectrum) and radar altimeter (for wave height) are techniques still under development and not yet readily available (refs. 11-12).

TIDAL ELEVATIONS AND CURRENTS

12. Tides are the periodic movements of the seas caused either by the regular movements of the Earth-Moon-Sun system (gravitational tides) or by periodic variations of atmospheric pressure and onshore-offshore winds (meteorological tides). Predictions of the tide are of relevance to coastal management because of their use for port operations, ship navigation, marine surveys and flood warning systems.

13. Tidal harmonic analysis of observed sea level data is carried out by the POL to yield the tidal "constants", amplitude and phase (time of arrival), of the constituent tidal waves, which are then used to compute tidal predictions.

14. Current studies in the theoretical and statistical analysis of tides (and surges) contribute to a better physical understanding of the complex mechanism of tide-tide and tide-surge interactions around the U.K. coast, which when incorporated into analysis and prediction techniques in turn helps to improve the quality of predictions (refs. 13-15), especially those of extreme tidal level (Fig. 3).

15. Two dimensional numerical models of tides have been developed for the U.K. continental shelf generally (refs. 16-18) and for more localised harbours and estuaries (refs. 19-20). The present thrust of research is concerned with better modelling of the physics (particularly frictional dissipation), smaller resolution models, and development of existing models for use on micro-computers. Three dimensional modelling, giving vertical profiles of current, of application for example in sedimentation studies, is an active research topic (ref. 21).

RESIDUAL ELEVATIONS AND CURRENTS

16. Residuals are the irregular non-tidal components of sea level or current which remain after removing the tides. In U.K. waters, the dominant contribution is from surges which are attributable to meteorological forcing of the sea surface by variations in atmospheric pressure or wind stress. The term "storm surge" is usually used to describe a specific storm event which generates large surge residuals. Such events are important in coastal management

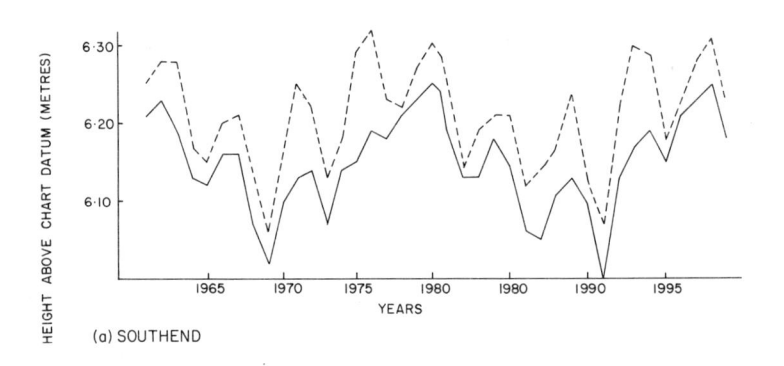

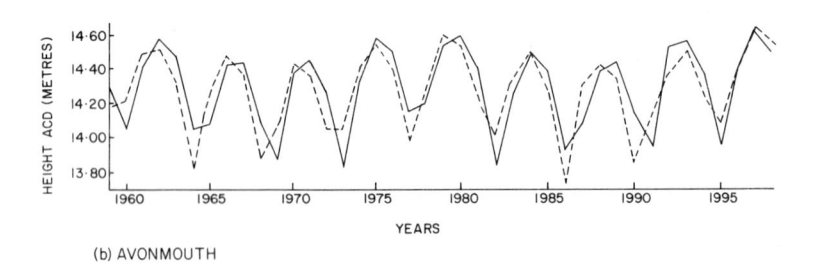

Fig. 3. Highest Spring and Autumn equinox tides.
───── Spring equinox
───── Autumn equinox

because of their effect on ship navigation, coastal structures and flooding potential.

17. Monthly and annual mean and standard deviation of surge elevation residuals are produced by the POL as part of its data processing, and can be used to generate the frequency distribution and amplitude and duration of extreme surges at a location.

18. Since 1978 a surge prediction model developed at the POL has been used at the Meteorological Office for operational storm surge elevation forecasting, using forecasts of wind and atmospheric pressure derived from a weather prediction model to compute surge sea levels and depth-averaged currents (ref. 22). During the "surge season" (September to March) two forecasts are produced every day, each one predicting the developing meteorological impact on the sea up to 36 hours ahead. Current work ensures that the model is maintained in good order to match evolution of the general computer hardware and software in use at the Meteorological Office and to monitor performance, identify shortcomings and, where necessary, introduce modifications to improve the results.

19. Recent monitoring of the performance has shown that, for 1987/88 storm surge season, the root mean square of the differences between observed and modelled hourly surge residuals was between 0.11 and 0.16m on the east coast, and between 0.09 and 0.24m on the west coast depending on location. The accuracy of the model forecasts depends on both the initial distribution of surge elevation and current and on the meteorological forcing applied. At present the initial fields are derived from a hindcast run based on "improved" meteorological data, no use being made of actual surge observations. However the feasibility of initialising the model forecast using real time observed surge data is being investigated.

20. The operational surge model is not yet satisfactory for the U.K. west coast, the most severe anomalies being found in the Bristol Channel area (ref. 23). The challenge is posed by the rapid and sensitive response of the Irish and Celtic Seas (being directly affected by approaching weather systems and near to semi-diurnal tidal resonance). Present research is directed to an improved theoretical understanding, to be achieved largely through experimental work with modified models and utilisation of on-shore and off-shore observations recently collected in the Bristol Channel area. This research should also aid the understanding of the local surge variations to be expected in other U.K. bays and estuaries.

21. Modelling of residual currents in more localised estuaries or harbours (e.g. the Clyde, Port Talbot) has also been carried out (refs. 20, 24-25).

RESEARCH

MEAN SEA LEVEL

22. Mean sea level (msl) is the average level of the sea. Daily, monthly and annual variations of msl are due to changes in atmospheric pressure, wind stress, water density and/or ocean circulation. Long term (secular) changes may also be caused by global changes in water volume (e.g. by the melting of grounded ice or sea water expansion by heating) and to more local or regional land movements.

23. U.K. msl data is collated at the POL, the data having been obtained by analysis of sea level records by Harbour Authorities, the MOD Hydrographic Department and by the POL itself. The Permanent Service for Mean Sea Level (PSMSL) is also situated at the POL and is responsible for the collection, publication and distribution of U.K. and world msl data; a catalogue of all data held is available on request.

24. Recent research at the POL has identified the spatial and temporal changes in U.K. msl (ref. 26); and shown that even in a relatively tectonically-stable area such as the British Isles, there are significant spatial variations of apparent msl rise due to vertical crustal movements. In the U.K., these vertical movements are of a similar magnitude to the true sea level rise and vary significantly spatially.

25. Advanced geodetic techniques (very long baseline interferometry (VLBI), Global Positioning Satellite (GPS) measurements and microgravity) have the potential to measure independently the vertical land movement of tide gauges, so that it can be removed from the tide gauge data to give real sea level changes (ref. 12). The Universities of Edinburgh and Nottingham, the Ordnance Survey and the POL are currently collaborating on a Geodesy "Special Topic" to make such measurements at some U.K. tide gauge sites.

26. Global sea level has risen by about 150mm over the last century and expectations are that it will increase appreciably in the next (perhaps by 300-700mm by 2050). Research at the POL (together with other European Laboratories including the Universities of Durham and East Anglia) is concerned with the secular msl changes on the European coast; particularly with the projected increasing rates of rise, possibly owing to the "greenhouse effect". A paper discussing the implications of sea level rise for coastal management is presented by Dr. Peerbolte in these proceedings.

EXTREME STILL WATER LEVEL

27. Still water level (swl) is defined as the level of the sea when waves have been averaged out. Extreme values of this parameter are of obvious use for coastal management purposes. Two methods are usually used to estimate extreme

swl: the Extreme Value (EV) method and the Combined or Joint Probability (JP) method.

28. The EV method produces estimates of extreme swl which are sensitive to the analysed data length, the inclusion or exclusion of particular values, and the method of curve fitting (ref. 27).

29. Since the introduction of the JP method (ref. 28), a number of deficiencies with the approach have been identified, concerned with the inadequacy of the empirical surge distribution and the non-independence of consecutive hourly sea level data, and its consequent effect on return period estimates. Both deficiencies especially lead to problems at sites which are surge-dominated, e.g. Lowestoft.

30. Recent progress has been made by developing the EV method to include a number of independent extreme values from each year of record (ref. 29), and by taking fuller account of dependence between hourly records in the JP method (ref. 30).

31. The modifications to both methods involve the use of a "storm length" to identify independent storms at a site and an "extremal index" to characterise the duration of a storm event. Consistent results are produced using the two revised methods (Fig. 4).

32. Further research is underway to handle tide-surge interaction and to extend the methods to adjacent sites with less data by transferring surge data from a neighbouring port with a longer data set.

WAVES

Offshore wave climate

33. The offshore wave climate can be determined by extrapolating directly from available wave data; relating available wave data to simultaneous wind data and using longer-established wind records for extrapolation; or using wave models in "hindcast" mode, driven by known past wind fields, to generate wave data additional to observations.

34. Lack of available wave data limits the reliability of the wave climate estimated by extrapolating an extreme value distribution fitted to the data (ref. 31). The few long data sets available indicate considerable interannual variation which makes extrapolation from a one year dataset dangerous (ref. 31). Long term trends in the data are not treated in present methods although recent studies indicate such non-stationarity (ref. 32).

35. Wave climate statistics can be synthesised from wind observations if both the conditional distribution of wave height relative to the wind and the long term distribution of wind are known. A method based on wind and global wind wave observations has been developed by the Meteorological Office and NMI Ltd. (now BMT Ltd.) (ref. 33) and is available commercially as a package on Personal Computers

RESEARCH

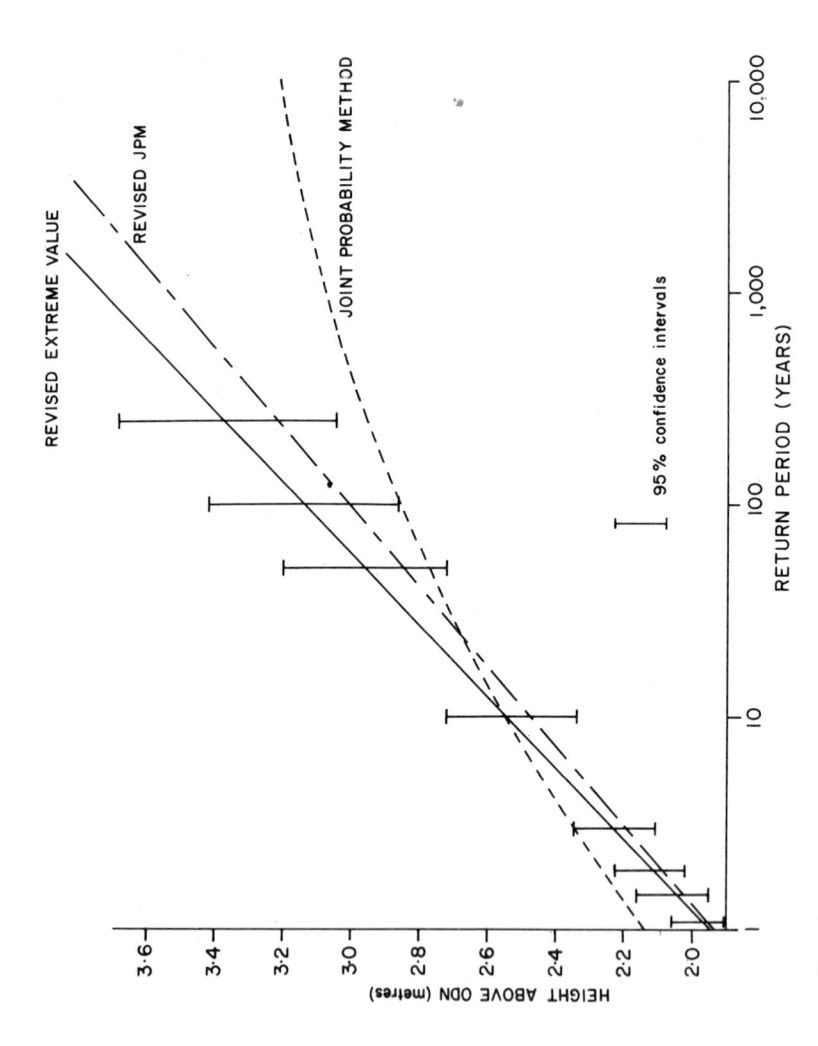

Fig. 4. Extreme levels at Lowestoft.

(PC) operating DOS.

36. Numerical methods of wave prediction are based on the integration of the energy balance equation for water waves, and the output is in the form of the frequency spectrum or possibly the complete directional spectrum for given times and selected grid points. The Meteorological Office wave model is run operationally for the British shelf seas, with wind data from weather stations and ships input at regular intervals (ref. 34). Forecasts are considered to be valid to an inshore depth of 10 metres but the grid resolution is only 25km.

37. Waves interact with and are modified by the changing currents and water levels associated with tides and surges, especially in shallow sea areas like the southern North Sea. The POL is developing a combined wave-surge-tide model, incorporating these interaction effects and based on the third-generation WAM wave model (refs. 35-36). The grid spacing is 10km over the continental shelf and 100 metres in selected coastal areas, with the model physics valid up to the breaker-zone. The combined model should overcome the difficulties of correcting existing separate wave, tide or surge models for these non-linear interactions.

Nearshore wave climate

38. As waves propagate inshore to depths less than half their wavelength, they become subject to the influence of the sea bed; effects include refraction, shoaling and dissipation through bottom friction, percolation, bottom motion and wave breaking. Changes of depth on scales comparable with or less than a wavelength, or barriers, sea walls etc., cause diffraction and reflection. Refraction and wave steepening may also result from spatially varying currents. Any or all of these effects may need to be taken into account in relating the waves inshore to those further offshore.

39. Theoretical and laboratory studies have recently been carried out on effects of non-linearity, wave breaking and wave-current interaction (refs. 37-42). The research is being used to develop techniques for calculating the wave behaviour in complex shallow water situations, and some has already been incorporated into wave-prediction programmes available commercially.

40. An example is a model originally developed at Liverpool University and available from Halcrow Ltd. As well as including refraction, reflection, diffraction and breaking effects, it also allows for the dynamic effects of a regular train of waves as they encounter harbour walls or breakwaters (personal communication, T. Hedges).

41. Some models use a wave tracing method to follow offshore waves as they approach the site and to calculate resulting changes in wave height and direction (ref. 43). In order to avoid the problem of merging or crossing rays

and of including rays which are refracted to an area outside the specific site of interest, Hydraulics Research Ltd. (HR) have developed a reverse-tracking method (ref. 44). Tracking starts at the site and follows a path out to sea, allowing for wave refraction and shoaling over varying bathymetry and also for effects of tidal currents.

42. HR have also developed a numerical procedure to hindcast hourly wave conditions at a point, from radial fetch lengths, measured wind records and the Jonswap forecasting method (ref. 44). The results build into monthly and overall histograms of wave height and direction, from which extrapolation to extremes can be carried out. The model estimates the mean annual wave climate, which can then be used to extend the short wave record to a longer, more representative period.

Wave set-up

43. The steady wave-induced increase in mean sea level produced shore-ward of the breaker line on beaches is called wave set-up, and may be a significant fraction of the wave height in deep water offshore (ref. 45). The POL and the IOSDL are currently carrying out research on wave set-up, and a set-up of 0.5m has been measured on Woolacombe beach (a beach of relatively gentle slope 1:70). At Chesil beach (slope about 1:7) 0.5m has been exceeded on several occasions with an average set-up over a two hour period of more than 1m during a storm in March 1987. The results from these studies have shown that the amount of set-up is strongly dependent on beach geometry and present work is concerned with extending the field experiments to adequately cover the range of beach geometry which occurs around the U.K. and to improve the theoretical studies.

WAVES AND STILL WATER LEVELS

44. In the U.K., post-1953 coastal defence design was based on a specified extreme still water level (swl) (often the 1953 observed level), and waves were allowed for by adding a certain additional free-board, based on local knowledge and experience. Following the series of coastal floods during 1976-78, more modern design thinking requires the consideration of the effect of combinations of swl and waves, since it is recognised that critical extreme events are not necessarily due to extreme events of either variable alone (ref. 27).

45. The extreme cases are that the swl and waves are either fully dependent or completely independent - in the first case the swl and wave height are added together; in the second the joint probability of the event is the product of the separate probabilities.

46. The design of proposed coastal defences or the assessment of existing ones therefore depends on estimating the joint probabilities of swl and waves, and the problem is to determine the degree of swl-wave correlation at a

particular site. This will vary from one coastal site to another depending in part on the exposure of the site to different wind directions. If the meteorological system generating the surges causes strong on-shore winds from a direction having a long fetch, then surges and waves are likely to be strongly dependent. If the local winds are off-shore, then surges and waves are likely to be less dependent (ref. 46).

47. Research into methods of combining the joint probabilities is currently underway at HR and the POL, but is hampered by the lack of even short term simultaneous swl and wave data sets. The methods have been applied to isolated sites where long term swl is available but where wave data has had to be synthesised from wind records (personal communication, M. Owen).

INTEGRATED STUDIES

48. Several studies have recently been started, for specific coastlines and estuaries, which draw together data gathering and computer modelling of the above, and related, topics. Examples are studies for South West Water of the Exe Estuary, the Taw-Torridge Estuary and Torbay; and a study for Anglian Water of its coastline. The studies are being carried out by various public and private sector organisations and companies, with overall project management by METOCEAN Ltd. and Halcrow Ltd. respectively. Such studies should provide a sound scientific base for effective coastal management strategies. A paper discussing the Anglian coastline project is presented by Dr. Fleming in these proceedings.

SUMMARY

49. Sea level data is adequately available from the modernised national network, supplemented by regional networks. Data on currents is only available as a result of specific coastal studies, but the measurement of coastal surface currents has been made more feasible by the development of OSCR; a miniaturised automatic system should make such measurements easier to obtain. In the future, satellite altimetry and SAR hold out the prospect of much improved global wave statistics, and the wave spectrum out to 100-200km offshore should be able to be inferred from shore-based high-frequency radar. However, the routine collection of wave data has actually been reduced since the ICE strong recommendation for a national plan for wave data acquisition. There is sparse wave data available now, and even less directional wave information, especially for validation of inshore wave models.

50. Tide and surge level and current modelling is well developed but ongoing studies aim to reduce west coast surge modelling uncertainties. Models are becoming more portable for running on PCs and therefore more available to

local engineers. Advances have been made in developing methods for estimating extreme still water levels due to tide and surge. Further work is needed to extend the methods to sites with short periods of available data. The greatest uncertainty in sea level studies lies in long-term projection of mean level changes. A prospective contribution is from the development of precise positioning techniques to enable separation of crustal movements from sea level changes in the relative measurements recorded by tide gauges.

51. Wave modelling is advancing in several respects, assisted by increasing computer power and theoretical laboratory experiments on the effects of non-linearity, wave breaking and wave-current interaction, especially nearshore. The development of a combined wave-surge-tide model incorporating non-linear interactions will be a significant advance. Methods of combining the joint probabilities of waves and still water level are currently being evolved but, like all research involving waves, is hampered by the lack of wave data.

52. There has been some development, in all research fields, of readily accessible low cost computational methods able to be employed by local engineers, but more of this "technology transfer" is presumably desirable.

53. Integrated studies of specific areas should provide a sound scientific base for effective coastal management.

ACKNOWLEDGEMENT

54. Much of the research described in this paper was, and is, funded by the Ministry of Agriculture, Fisheries and Food, the Natural Environment Research Council and the Science Engineering Research Council.

REFERENCES

1. INSTITUTE OF CIVIL ENGINEERS. Research requirements in Coastal Engineering - a report prepared by the Coastal Engineering Research Panel and Working Parties, 123pp. Thomas Telford, London, 1985.
2. RAE J.B. Development and Operation of the U.K. Tide Gauge Network. MAFF Conference of River Engineers, July 1986, Cranfield.
3. ALCOCK G. The IOS sea level data collection. The Dock and Harbour Authority, 1985, vol. 65, 121-123.
4. BUCKLEY K. Tidal Monitoring and Issue of Warnings, Anglian Water Norwich Division. MAFF Conference of River Engineers, July 1986, Cranfield.
5. FLEMING J.H., McMILLAN P.H. and WILLIAMS B.P. The River Hull tidal surge barrier. Proceedings of the Institution of Civil Engineers 1980, vol. 68, 417-454. Discussion: Ibid, vol. 70, 1981, 581-591.
6. HILLIER G.J. and MILLER K. A new tidal data acquisition and processing system for the Thames. The Dock and Harbour Authority, 1987, vol. 68 (791), 33-35.

7. BARBER P.C. and DAVIES C.D. Offshore breakwaters - Leasowe Bay. Proceedings of the Institution of Civil Engineers 1985, vol. 78, 85-109. Discussion: Ibid, vol. 80, 1579-1608.
8. PRANDLE D. A review of the use of HF radar (OSCR) for measuring near-shore surface currents. Proceedings of Second Joint International Symposium on Water Modelling and Measurement, Harrogate, April 1989.
9. COLLAR P.G. and HOWARTH M.J. An intercomparison of currents measured by OSCR, current meters and drogued floats within one metre of the sea surface. Department of Energy, Offshore Technology Report OTH 87 272, HMSO, London, 1987.
10. THORNE K.L. and GLEASON R. Waves recorded off Kinnairds Head. Institute of Oceanographic Sciences Report no. 226, 1986, 69p.
11. WHYATT L.R., VENN J. and BURROWS G.D. Ocean surface wave and current measurement with HF ground-wave radar. pp. 15-25 in, International conference on measuring techniques of hydraulics phenomena in offshore, coastal and inland waters, London. Cranfield, Bedford: BHRA, 1986, 451p.
12. HUTHNANCE J.M. Review of the potential of satellite remote sensing for marine flood protection. Institute of Oceanographic Sciences, 1987, Report No. 237.
13. AMIN M. Temporal variations of tides on the west coast of Great Britain. Geophysical Journal of the Royal Astronomical Society, 1985, vol. 82, 279-299.
14. AMIN M. "On the conditions for classification of tides". International Hydrographic Review, 1986, vol. 63(1), 161-174.
15. AMIN M. A method for approximating the nodal modulations of real tides. International Hydrographic Review, 1987, vol. 64(2), 103-113.
16. FLATHER R.A. A tidal model of the North-West European continental shelf. Memoires de la Societe Royal des Sciences de Liege 1976, vol. 10, 141-162.
17. PINGREE R.D. and GRIFFITHS D.K. S_2 tidal simulations on the north-west European shelf. Journal of the Marine Biological Association of the U.K. 1981, vol. 61, 609-616.
18. PRANDLE D. Co-tidal charts for the Southern North Sea. Deutsche Hydrographische Zeitschrift, 1980, vol. 33, 68-81.
19. FALCONER R.A. and OWENS P.H. Numerical simulation of flooding and drying in a depth-averaged tidal flow model. Proceedings of the Institution of Civil Engineers, Part 2, 1987, vol. 83, 161-180.
20. UNCLES R.J., JORDAN M.B. and TAYLOR A.H. Temporal variability of elevations, currents and salinity in a well-mixed estuary. pp. 103-122 in, Estuarine variability (ed. D.A. Wolfe), Academic Press, New York, 1986.

21. DAVIES A.M. A three-dimensional model of the northwest European continental shelf with application to the M. tide. Journal of Physical Oceanography, 1986, vol. 16, 797-813.
22. FLATHER R.A. Practical surge prediction using numerical models. pp. 21-43 in Floods due to high winds and tides. (Ed. D.H. Peregrine). Academic Press, London, 1981.
23. PROCTOR R. and FLATHER R.A. Storm surge prediction in the Bristol Channel - the floods of 13th December 1981. 1989, submitted to Continental Shelf Research.
24. FALCONER R.A. Residual currents in Port Talbort Harbour: a mathematical model study. Proceedings of the Institution of Civil Engineers, Part 2, 1985, vol. 79, 33-53.
25. TOWNSON J.M. and DONALD A.S. Numerical modelling of storm surges in the Clyde Sea area. Proceedings of the Institution of Civil Engineers, 1985, vol. 79, 637-655.
26. WOODWORTH P.L. Trends in U.K. mean sea level. Marine Geodesy, 1987, vol. 11, 57-87.
27. ALCOCK G. Parameterizing extreme still water levels and waves in design level studies. Institute of oceanographic Sciences Report, 1984, No. 183, 95p.
28. PUGH D.T. and VASSIE J.M. Applications of the joint probability method for extreme sea level computations. Proceedings of the Institution of Civil Engineers, 1980, vol. 69, 959-975.
29. TAWN J.A. and VASSIE J.M. The joint probability method revisited and revised. Proceedings of the Institution of Civil Engineers, 1989, (in press).
30. TAWN J.A. Extreme value theory model for dependent observations, Journal of Hydrology, 1988, vol. 101, 227-250.
31. CARTER D.J.T. and TUCKER M.J. Uncertainties in Environmental Design Criteria. Underwater Technology, 1986, vol. 12, no. 1, 2-8.
32. CARTER D.J.T. and DRAPER L. Has the north-east Atlantic become rougher? Nature, 1988, vol. 332 (6164), p 494.
33. HOGBEN N. Experience from compilation of global wave statistics. Ocean Engineering, 1988, vol. 15, 1-31.
34. FRANCIS P.E. Numerical wave models as a source of data for marine climatology. In: Advances in underwater technology, ocean science and offshore engineering, vol. 6, 157-166. Graham and Trotman, London, 1986.
35. KOMEN G.J. Activities of the WAM (Wave Modelling) Group. In: Advances in underwater technology, ocean science and offshore engineering, vol. 6, 121-127. Graham and Trotman, London, 1986.
36. WOLF J., HUBBERT K.P. and FLATHER R.A. A feasibility study for the development of a joint surge and wave model. Proudman Oceanographic Laboratory Report No. 1, 109p.

37. PEREGRINE D.H. Non-linear wave refraction. University of Bristol School of Mathematics, report no. AM-88-06, 1988, 31p.

38. PEREGRINE D.H. Breaking water waves. University of Bristol School of Mathematics, 1988, report no. AM-88-07, 37pp.

39. YOO D., HEDGES T.S. and O'CONNOR B.A. Numerical modelling of reflective waves on slowly-varying currents. Proceedings of Congress on Water Modelling and Measurement, IAHR-BHRA, 1988.

40. ANASTASIOU K. Experimental investigation of the kinematics of breaking waves. Research in the Dept. of Civil Engineering, Imperial College of Science and Technology. (Personal communication from Dr. E. Forster, MTD Ltd.)

41. HYDRAULICS RESEARCH LTD. Current-depth refraction of water waves, A description and verification of three numerical models. HR Report No. SR14, 1985.

42. SIMONS R.R. and GRASS A.J. Wave attenuation in the presence of coastal currents. Ongoing research in the Department of Civil Engineering, University College, London. (Personal communication from Dr. R. Thorogood, SERC.)

43. ELLIOTT A.J. Wave predictions by an empirical ray tracing model. Department of Energy, Offshore Technology Information OTI 87 502, HMSO, London, 1987.

44. HYDRAULICS RESEARCH LTD. Wave prediction in deep water and at the coastline. HR Report No. SR114, 1987.

45. JAMES I.D. The effects of wind waves on the sea level at the coast. Institute of Oceanographic Sciences Report No. 155, 66p.

46. ALCOCK G.A. and CARTER D.J.T. Extreme events in waves and water levels. In Developments in Breakwaters, Proceedings of Conference, Institution of Civil Engineers, London, 1985, 53-68.

8. Research on beaches and coastal structures

M. W. OWEN, MA, MICE, Research Manager, Coastal Engineering Group, Hydraulics Research Ltd

SYNOPSIS. The art of good management could be defined as "making the best possible use of the resources available in order to achieve specified objectives". In the field of coastal engineering, the usual objectives are either to minimise flooding or to reduce erosion due to the action of waves and tides. For these purposes the resources which are available obviously include the so called hard or traditional sea defence works such as seawalls and revetments. However in the past decade or so it has increasingly been recognised that natural or soft defences such as sand or shingle beaches, dune belts, and saltings are also a very valuable coastal defence resource. This paper describes some of the research presently being carried out on beaches and coastal structures, which will hopefully lead to improved management of sea defences.

BEACHES
1. The most important distinction between coastal structures and beaches is that beaches are continually changing, in timescales varying from a few minutes to a few months. Hopefully in the long term the destructive and the constructive forces on the beach will balance each other out, but often man has to intervene to assist the achievement of this state of dynamic equilibrium.

2. For the sucessful use of beaches as a means of coast defence, the design requirements are therefore:

a) Identification of the minimum beach profile necessary to give the required protection at all times.
b) Identification of the plan shape necessary to give the required beach profile at all important locations along the coast.
c) Evaluation of any beach structures or management schemes which are necessary to maintain the required beach shape.

Beach profile
3. The minimum beach profile required to protect the coastline during major storms clearly depends on the sediment size and grading and on tide levels and wave conditions. Probably the most accurate way of determining the profile is by

RESEARCH

very careful monitoring of a nearby beach having a similar material, and the same exposure. Very many coastal engineers carry out regular surveys of the beaches for which they are responsible, but often very little analysis is performed on the results obtained. However micro-computer software packages are now becoming available which are used to archive beach cross-sections in a standard format and to provide rapid recall of any particular section or groups of sections. More importantly, the cross sectional data can be analysed in a variety of ways, depending on the number and frequency of the surveys. Both fluctuations and long term trends in beach levels are very important in considering the state of the beach. For example, a sudden drop in beach levels as a result of a major storm can be put into its proper perspective by comparing it with past beach level fluctuations. Conversely, a good historical data set indicating a continuing trend of beach erosion will make it much easier to argue the case for an engineering scheme.

4. As indicated, good historical data is the best guide for a coast defence scheme. Often however design has to proceed without such data, either because it is unavailable, or because there is no similar beach nearby. Fortunately there are now some reasonably reliable predictive techniques available, especially for shingle beaches. Firstly a scale model could be constructed, in which the model sediments should be carefully selected to reproduce the movement of the individual particles, and the permeability of the resulting beach. Although there are still some difficulties in scaling sand, shingle beaches can be modelled quite accurately, and several major shingle beach nourishment schemes in the UK have been modelled recently (Refs 1, 2). However not every scheme can justify the major expense of model studies. Since 1984 therefore a major research project on beaches has been underway at Hydraulics Research Ltd : to date this has concentrated entirely on shingle beaches and mostly on waves striking the beach at normal incidence. This research has been based on a very extensive and systematic series of random wave, mobile bed, model tests with varying wave height, wave period, sea steepness, storm duration, shingle size and degree of sorting. In addition, some tests were carried out with varying depths of shingle over an impermeable basement. During each test, measurements were made of wave run-up levels, reflection coefficients and beach profiles.

5. As can be imagined, this research has produced a considerable quantity of results to be analysed. It has been found that the beach profile can be characterised by three distinct curves, of the type shown in Fig 1. The characteristic profile therefore contains nine predictors, and from the model results expressions have been derived which relate the value of each predictor to the input variables. All these expressions are given in the research report to be published shortly (Ref 3), and from these it should be possible to predict the complete beach profile for any given input conditions, although for normal wave attack only at present. One interesting result

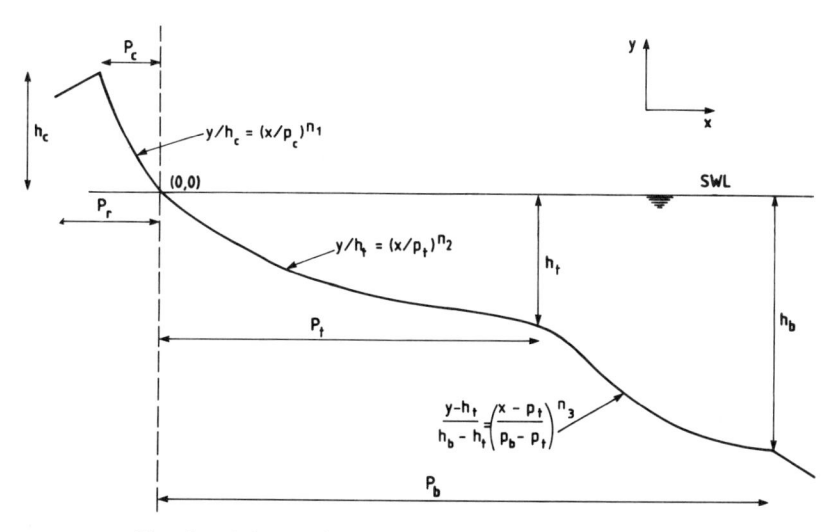

Fig 1 Schematisation of shingle beach profile

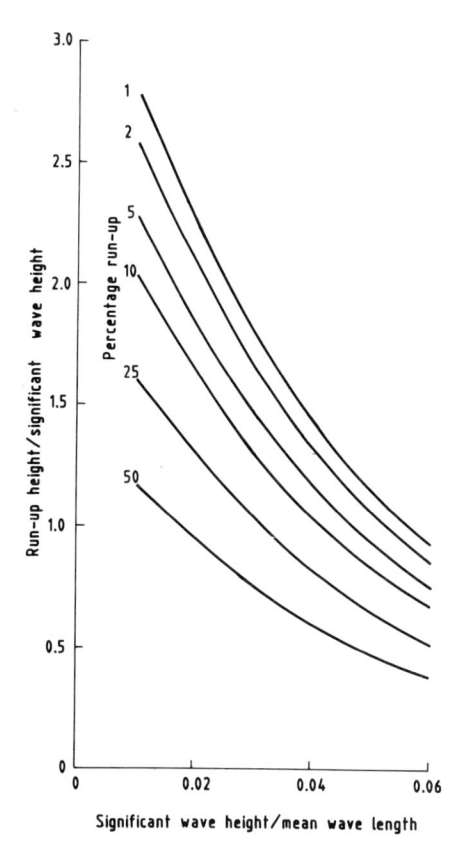

Fig 2 Run-up on shingle beaches

which emerged from the model studies was that the height of the beach crest and the level of wave run-up on the beach depend only on the wave height and wave period. Consequently, the run-up level for example can be represented in a very simple graph, Fig 2. This graph can also be used to obtain a first estimate of the height of the beach crest, since in most tests the crest height was very similar to the run-up level achieved by about 1-3% of the waves.

Plan shape
6. The required plan shape of the beach has to ensure that the minimum beach profile is equalled or exceeded at all important locations along the coastline. Since the plan shape on some coasts can change quite rapidly from storm to storm, then obviously it is necessary to be able to predict those changes, to ensure that the necessary protection is obtained at all times. Again, it is often possible to deduce the required plan shape from a detailed examination of historical data on nearby beaches. However, such data does not usually indicate the short term re-alignment of the beach during specific storms.

7. Traditionally the plan shape of beaches has been determined from mobile bed model tests. Such models are still used today, especially where the existing coastline has a fairly complex shape, or where the inshore bathymetry is very variable. However in very many cases, mathematical modelling techniques are used to predict beach plan shape. The simplest of these models were developed about 15 years ago (Ref 4). The models assume that the beach profile and the water level stay constant at all times, and the plan shape of the beach can therefore be represented by the position of a single contour. These one-line models are very simple and it is possible to contemplate running them for a few hundred wave conditions each year, and for several decades, to determine how the beach plan shape changes from day to day, and how it develops in the long term.

8. The original one-line models could deal only with sand beaches, and only with very simple coastal structures. Recently however, further research has enabled one-line models to be extended to beaches where there is a marked change in wave height along the coastline, to beaches backed by seawalls (Ref 5), and to shingle beaches (Ref 6). In certain circumstances more complex beach structures can also be introduced into one-line models, but these are described in the next section.

Beach structures and management schemes
9. It is often not possible to maintain the required plan shape for the beach without some positive action. Usually this involves one of three options:
a) Regularly import beach materials to replace the quantities lost (regular renourishment).
b) Trap the beach materials at the downdrift end, perhaps by constructing a very large groyne. Remove the materials

from this area, and place them at the updrift end (recycling).

c) Construct and maintain a series of beach structures, such as groynes and offshore breakwaters, to reduce or possibly eliminate the alongshore drift of beach materials.

10. The selection of the optimum scheme depends critically on the ability to predict the rate of alongshore transport of beach materials, and also the effectiveness of any proposed beach structures. The regular costs of recycling or renourishment can then be calculated, and compared with the cost of building and maintaining the beach structures. For regular renourishment and recycling schemes, the simple one-line model can give reasonably accurate estimates of the quantities of beach sediment which have to be moved, depending both on the wave climate and on the frequency of the operations. The model also costs a relatively modest amount. The main difficulty comes in predicting the long term effectiveness of the proposed structures. As a result of recent research, mathematical models with or without complementary physical models are now usually employed for this purpose. On the whole, physical models give a better representation of detailed beach changes close to coastal structures, or on coastlines of rather complex shape. However, because they are fairly expensive to operate, usually only a few wave conditions and water levels can economically be examined. As mentioned earlier, one-line beach mathematical models can cover many hundreds of wave conditions, but only for fairly straightforward coastlines or simple beach structures. For more complex structures, such as short groynes, or offshore breakwaters, the one-line model has to be provided with input on the behaviour of the structure.

11. As an example, for the studies of the coast protection works at Hengistbury Head, near Bournemouth (Ref 7), a mobile bed random wave physical model was constructed, mainly because of the rather complex coastline in that area. However, this model was also used to measure the bypass rates for both sand and shingle around a typical rock groyne of various lengths, for a few wave conditions. This information was then used as input to a simple beach mathematical model of the coastline. Fig 3 shows the coastline east of Hengistbury Head Long Groyne, and indicates the predicted positions of the beaches within each groyne compartment 5 years after placing nourishment material along the Hengistbury Head frontage (sections 46 - 76).

12. Clearly, for many schemes the cost of a large physical model is not justified, where a mathematical model would be acceptable. For schemes which do involve fairly complex beach structures, there are two ways forward. One would be to carry out some systematic research on the behaviour of such structures, to provide all the necessary functions for the mathematical model. The other is to develop more complex mathematical models, in which the various physical processes are better represented. The recent CIRIA research project on the effectiveness of groyne systems is an example of the former

RESEARCH

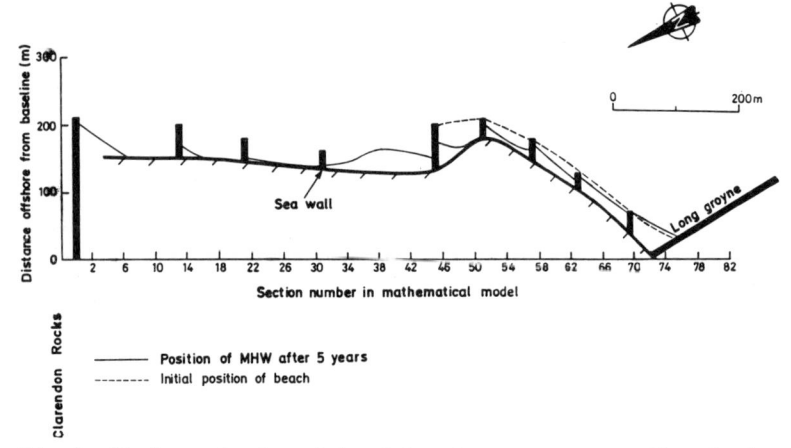

Fig 3 Mathematical model of beach movements at Hengistbury Head

approach (Ref 8). However, to everyone's disappointment the various studies which were carried out did not enable any firm recommendations to be made on the design of new groyne systems, especially for sandy beaches. More work on this topic seems to be necessary, probably including systematic mobile bed physical model testing.

13. The development of more realistic mathematical models is proceeding in many countries. Ideally, such models should marry together the equations of water motion on the beach with those of sediment transport. Ultimately, these models will be able to perform more accurately all the tasks which presently require both physical and simple mathematical models. At the moment, these morphodynamic models are hampered by the lack of suitable equations describing accurately the complex interactions between waves, currents and sediment transport in the highly turbulent surf zone. Although a few such models have been attempted, their very high costs and doubtful accuracy make their use difficult to justify at present for designing engineering schemes. Most of the research effort is therefore being directed at obtaining a better understanding of the various physical processes involved.

14. In the meantime, the multi-line beach mathematical model can be used (Ref 9), which while not able to tackle the most complex beach structures nevertheless offers several advantages over the simple one-line model. The multi-line model includes both alongshore and onshore/offshore transport, and allows the beach profile to deviate from its long term average. The gross alongshore transport is distributed across the beach width, depending on the wave height and the position of the breaker line, and the position of every beach contour is predicted. Because of its ability to include different transport rates on different parts of the beach, the multi-line model can be used to study mixed sand and shingle beaches or to examine the

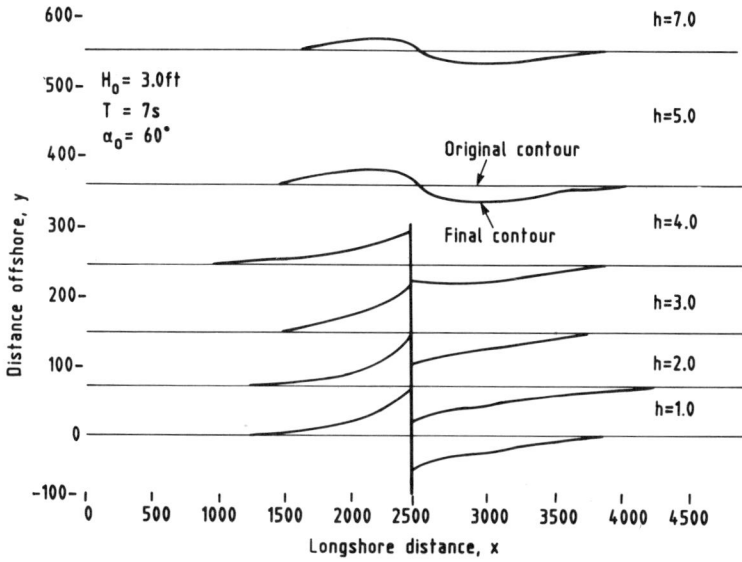

Fig 4. Multi-line model of beach changes at a groyne (after Ref 9)

effectiveness of short groynes. Fig 4 for example shows the beach contours around such a groyne: for those contours seaward of the groyne tip, the longshore transport rate is assumed to be unchanged, whereas on the contours landward of the tip the longshore transport is assumed to be totally arrested by the groyne. Fig 4 is for a single combination of wave height, period and direction, and for a single water level. However, the model can be run for many different combinations, each having a different longshore transport rate past the groyne. By summation, the overall effectiveness of the groyne can then be determined.

Future research requirements

15. In order to make much better use of the mathematical modelling techniques which are currently practicable, a series of systematic model studies and field validation measurements is necessary to parameterise the profiles of both sand and shingle beaches under angled wave attack, and also to derive empirical expressions describing the effectiveness of complex beach structures. In the long term however, a comprehensive research programme is needed which will lead to a full coastal morphodynamic model, in which all processes of water motion and sediment transport in the surf zone are represented. This major task is likely to require international co-operation to achieve, but once available the model should be able to tackle almost any coastal defence scheme involving beaches.

COASTAL STRUCTURES

16. The term "coastal structures" can include a whole host of artefacts, but in this paper it is taken to refer only to sea defence works such as seawalls, revetments and offshore (shore

RESEARCH

parallel) breakwaters. In the design of such structures, the two main requirements are:

a) Identification of the hydraulic profile necessary to give the required protection
b) Derivation of a structural form which will remain stable under the design conditions.

Hydraulic profile

17. For seawalls and revetments, the main requirement is to limit wave run-up and overtopping under the design storm conditions. The literature review which was published recently as part of the CIRIA Research Project on seawalls (Ref 10) showed that for smooth faced seawalls of fairly simple shape quite a lot of information was already available. However, there are very many other types of seawalls and revetments, including those incorporating rough and/or porous faces, recurved wave return walls, compound slopes etc, for which physical model studies are still necessary to determine the overtopping discharges. Shortly the results of research studies on some of these more complex seawalls should become available, allowing calculations to be carried out for overtopping discharges, at least for preliminary design purposes. The research is built upon the previous work on embankment - type seawalls (Ref 11), and the model test programme includes measurements of wave run-up and overtopping on seawalls having rough but impervious faces, on rubble-mound revetments, and on seawalls incorporating a recurved wave return wall at their crest. All measurements were carried out in an random-wave flume, and therefore assume normal wave attack. The test programme has now been completed but the many test results are still being analysed.

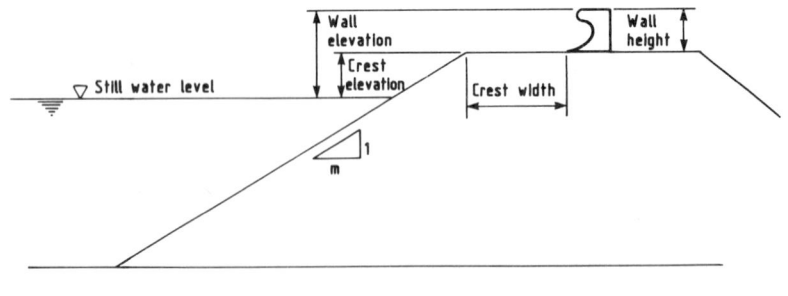

Crest elevations 0.5 1.0 1.5 m   Crest widths 0 4 8 m
Wall elevations 1.0 1.5 2.0 m   Main wall slopes 1:2 1:4

Fig 5 Generalised form of recurved wall test sections

18. As an example, some of the results of the tests with recurved wave return wall can be discussed. In these tests, a plain sloping embankment of fixed crest elevation was surmounted by recurved walls of three different heights, each placed at three different positions on the crest berm, Fig 5. The exact shape of each recurved wall was taken from the table originally derived by

Berkeley Thorn, and quoted by Owen (Ref 11). For recurved walls placed immediately at the top of the slope, Fig 6 shows a plot of the overtopping discharge against the total height of the seawall, for a particular wave condition and water level. The solid lines show the reduction in discharge which is obtained by increasing the height of the wave return wall, founded on a fixed embankment

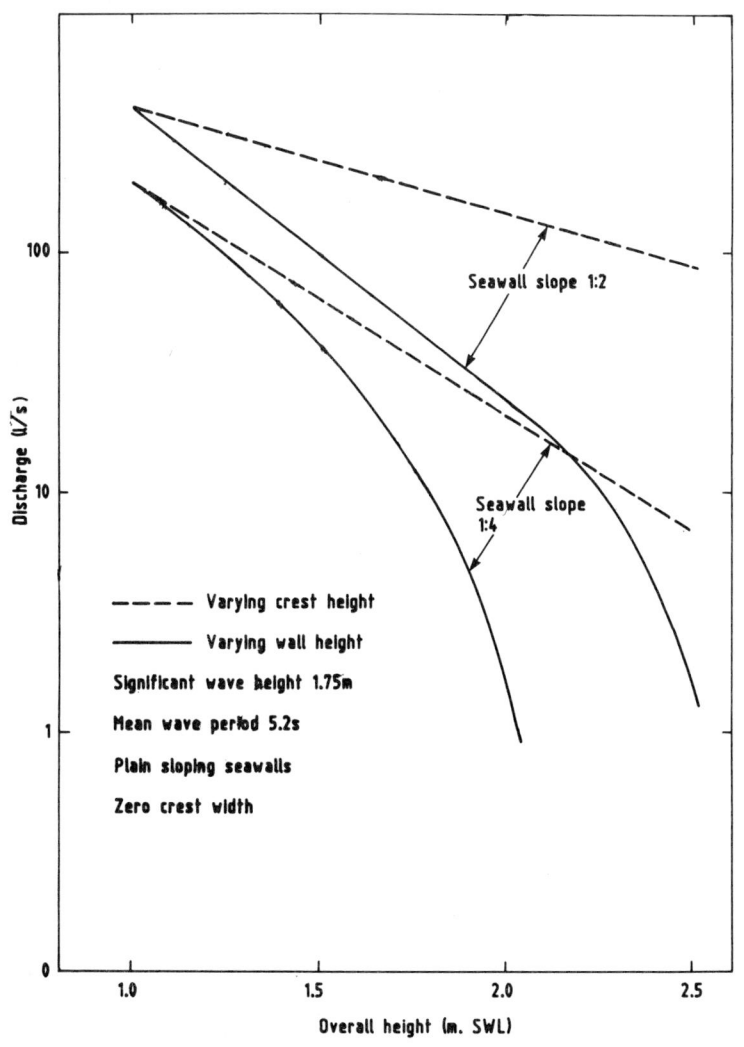

Fig 6   Effect of recurved walls on overtopping discharges

crest height. The dotted line shows the reduction obtained by raising the embankment crest height, without any wave return wall. For a given total height of seawall, the incorporation of a wave return wall greatly reduces the overtopping discharge, especially when the height of the return wall approaches the wave height.

RESEARCH

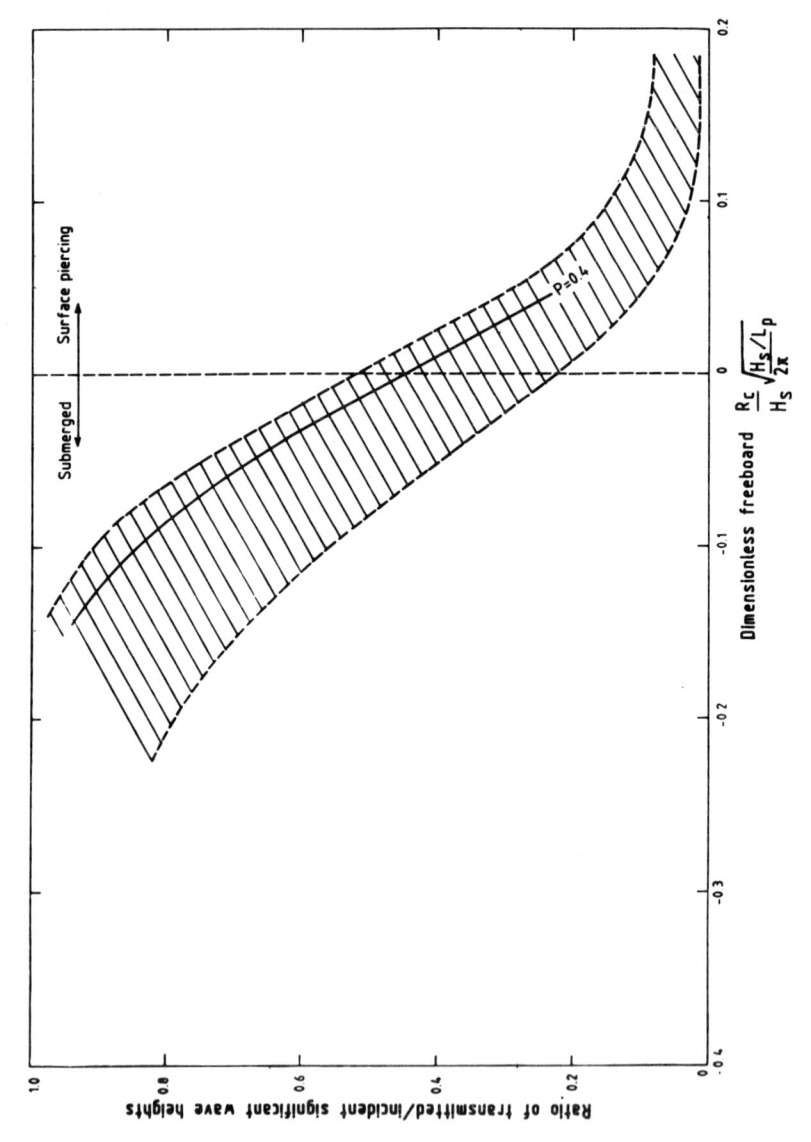

Fig 7 Wave transmission at low-crest rubble-mound breakwaters

19. For offshore breakwaters used as coastal defences the main requirement is to reduce the wave height at the shoreline by a specified amount, and over a specified distance. At present there is very little guidance available on the height, length, distance offshore, and spacing of these structures. Because offshore breakwaters are relatively short, predicting the wave conditions at the shoreline is essentially a three-dimensional problem. The waves derive from two sources - firstly diffraction around the ends of the breakwaters, and secondly waves transmitted either over or through the breakwater (Ref 12). Because the breakwaters are usually fairly close inshore, in an area where water depths are decreasing quite rapidly, accurate prediction of the heights of the diffracted waves as they penetrate into the lee of the breakwater is not at all straightforward. Normally a fairly complex site-specific mathematical model will be required, or else a physical model in a random wave basin.

20. The height of the waves transmitted over or through the breakwater will clearly depend on the form of construction, and on its height relative to the water level and wave height. Some guidance is available in the literature : Fig 7 for example shows design curves derived from a series of tests in random wave flumes for rubble mound breakwaters having seaward slopes in the range 1:1½ to 1:3 (Ref 13). The envelope of curves covers the range of likely breakwater porosities, ranging from about 0.5 to zero: a porosity of about 0.4 is common for most offshore breakwaters. It is important to realise that since the transmitted waves have a relatively short crest length (approximately equal to the breakwater length), then the transmitted waves will themselves be diffracted away from the area in the lee of the breakwater. At any position on the shoreline the total wave energy will therefore be the summation of the diffracted incident wave and the diffracted transmitted wave.

Structural stability

21. There is an enormous variety of structural form amongst seawalls and revetments, ranging through mass concrete, concrete blocks and articulated mattresses, to rubble mounds protected either by rock or a large range of different concrete armour units. However at the present time most attention is focussed on the so-called flexible revetments (articulated mattresses, open stone asphalt etc) and on rubble slopes protected by rock. Flexible revetments have been used quite extensively in inland waterways (Ref 14), but their use so far in coastal situations has been restricted to areas of fairly limited wave height. At the present time PIANC have an international working group preparing design guidelines for the use of flexible revetments in the marine environment: however it seems likely that there will be no firm recommendations for example on the required thickness of the cover layers, and its variation with porosity, revetment slope and wave conditions. Therefore, although there is a strong desire to use flexible revetments in more exposed locations, the problems of design are still not resolved, and there is at present very little research being carried out on this topic. Nevertheless an open

RESEARCH

stone asphalt seawall has recently been constructed at a fairly exposed location on the North Wales coast, and a research contract has been awarded jointly to the Polytechnic of Wales and Hydraulics Research Ltd to monitor its performance, and in particular to measure the attenuation of wave pressures through the 700mm thick cover layer. This work has only just started, and the results will be awaited with interest.

22. The use of rock in coastal structures has increased noticeably in recent years, partly as a result of new sources of good quality rock at reasonable cost from Scandinavia and from Spain (Ref 15). This increased interest has prompted CIRIA and the similar Dutch organisation CUR to begin work on a joint research project aimed at producing guidelines for the design, construction and maintenance of rock structures at the coastline. In the design phase, the guidelines will rely heavily on the research by van der Meer (Ref 16). This work was itself based on earlier studies carried out on CIRIA's behalf at Wallingford, and it is likely that some further tests on a limited scale will be carried out to investigate the effects of different shapes of size grading curve on the stability of the rock armour layer. In the pre-construction phase, the selection of suitable rock will be based mainly on the research carried out jointly by Queen Mary College, London and Hydraulics Research Ltd on the durability of rock in the marine environment. This research has resulted in a set of standard tests for rock quality, together with some recommended values for acceptance or rejection of the particular rock (Ref 15). These values are given below:

Density	minimum 2600 kg/m^3
Water absorption	maximum 2.5%
Magnesium sulphate soundness loss	maximum 12%
Aggregate impact value	maximum 25%
Franklin point load	minimum 4 MN/m^2
Fracture toughness	minimum 0.7 MN/m$^{3/2}$

23. As well as seawalls and revetments, rock can of course also be used to construct offshore breakwaters, and general harbour breakwaters. These breakwaters may also be armoured with concrete units such as dolosse, tetrapods, stabits, which are not very often used for revetment construction. However more and more attention has been focussed lately on single layer, high porosity concrete armour units, especially in the UK where COBS, SHEDS, and Diode units originated. A major research project is presently underway to investigate the performance of these units, under the auspices of a research club including academic institutions, research organisations, consulting engineers, contractors and breakwater owners. The research involves laboratory studies, field measurements, and mathematical modelling to investigate not only the hydraulic performance and stability of the units, but also the structural strength both of the individual units and of the complete armour layer. The research studies will hopefully lead to more efficient design procedures for such units, and if

this is achieved their use will probably spread to smaller structures such as coast defences and revetments. In addition, much of the expertise developed during the course of the research programme is likely to be of direct benefit in trying to derive more rigorous design methods for the use of flexible revetments in exposed coastal locations.

Future research requirements

24. In terms of the hydraulic profile of coastal structures, probably the most important item still lacking is an understanding of the effect of wave direction on run-up, overtopping and wave transmission. This can only be achieved by a systematic series of model tests, if possible calibrated against field measurements. For structural stability there are still a great many problems to be solved, including the determination of wave pressures on seawalls, the prediction of the performance of flexible revetments, and the effects of wave direction on rock armour stability etc.

INTERACTION BETWEEN STRUCTURES AND BEACHES

25. It is very rare that a coastal defence system consists entirely of structures or of beaches: in almost every case there is some element of each present. The sea defences can affect the beach in two ways, either by altering the balance of onshore and offshore transport, or by causing a discontinuity in the alongshore transport rate.

Onshore/offshore transport

26. At any position on most beaches there is a long term balance between the quantities of sediment moved onshore or offshore during different wave and tidal conditions. As a result, the beach profile is usually stable in the long term. When a seawall or revetment is constructed, a certain proportion of the incident wave energy is reflected back over the beach, causing a local imbalance in onshore/offshore transport, and hence altering the beach profile. Similarly an offshore breakwater may alter the transport balance and hence the beach profile. The general view is that seawalls and beaches cause beach erosion, while offshore breakwaters cause beach accretion. However, this is a considerable over-simplification because all coastal structures produce zones of both accretion and erosion. The problem is to predict the extent and magnitude of these changes.

27. At present, a major research programme is underway at Hydraulics Research Ltd to study the protential erosion of shingle beaches fronting seawalls. The research involves mobile bed model studies in a random wave flume, in which the beach profiles in front of the seawall are very carefully monitored for various designs and positions of seawalls. Vertical seawalls, sloping revetments, and rock mounds have been examined, at positions varying from the toe to the crest of the natural beach. The tests are now almost complete and the results are presently being analysed. Fig 8 shows a summary of the results obtained for

RESEARCH

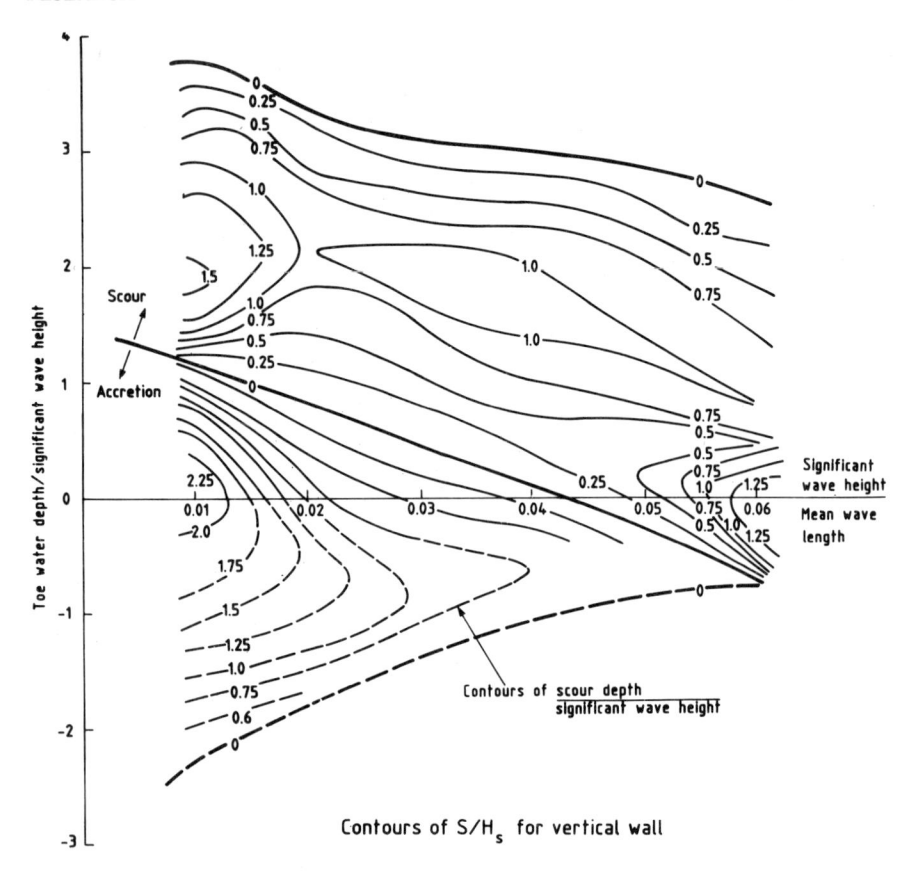

Fig 8 Shingle beach scour at the toe of vertical seawalls

vertical seawalls. The contours on this diagram represent the
dimensionless depth of scour immediately ajacent to the seawall.
This scour depth is a complex function of the wave steepness (wave
height/wave length), and of the dimensionless water depth at the
location of the seawall for the natural beach. For seawalls which
are located close to or above the still water line the beach level
actually builds up, except during periods of very steep storm
waves. However, for seawalls located below the still water line
erosion occurs for most wave conditions. In reality of course the
wave conditions are changing continually, and the net effect on
beach levels will be the summation of all the different erosion
and accretion depths. When the results of all the tests have been
analysed it will hopefully be possible to compare the scour
potential of different designs of seawall, and to predict the
exact shape of the beach profile after the seawall is
constructed.

Alongshore transport
  28. Because seawalls and revetments reflect wave energy, it
follows that the energy available to transport sediment alongshore

is reduced. If the beach profile in front of the seawall erodes to such an extent that the waves no longer break on the beach, then alongshore transport will cease completely. If the seawall has a finite length, then the effect of a seawall must be very similar to a terminal groyne, with very rapid downdrift erosion. According to simple theory, as a corollary there should also be an accretion zone near the updrift end of the seawall, but this is hardly ever noticeable, probably because the reflections off the seawall carry the material offshore. At present, there is very little firm guidance on the effect of seawalls on alongshore transport rates: where this is likely to pose problems a mobile bed physical model study would be necessary, probably combined with a simple one-line beach mathematical model. The results from a few tests in the physical model would be used to calibrate a purely empirical expression built into the mathematical model, which would then be used to examine very many more wave conditions. It would clearly be an advantage to have a much better understanding of the precise effects of such seawalls, so that a physically realistic expression could be used in the mathematical models.

29. For offshore breakwaters the effect on the beaches is due to changes in both alongshore and onshore/offshore transport and to the wave-induced current patterns which are formed. At present there is no straightforward way of predicting the resulting shape of the beach. Accurate prediction can only be achieved through model bed physical models, or through the development of the full morphodynamic mathematical models. Some of the present problems with such models were discussed in paragraph 14.

Future research requirements

30. All the work so far carried out on the scour of beaches fronting seawalls and revetments has been for normal wave attack. There is clearly a need to extend this work to examine angled wave attack, to determine the effects of the structures not only on beach profiles but also on alongshore transport rate, and its distribution across the surf zone. The need for a full coastal morphodynamic model to predict the performance of offshore breakwaters in retaining a beach has been mentioned previously.

AKNOWLEDGEMENTS

31. Most of the research studies described in this paper have been funded by the Ministry of Agriculture, Fisheries and Food, and/or by the Department of Environment (Construction Industry Directorate).

REFERENCES

1   Whitstable sea defences. Mobile bed model studies. Report No. EX 1355, Hydraulics Research Ltd, Wallingford, April 1985.
2   Seaford frontage study. Physical and numerical modelling, Report No. EX 1346, Hydraulics Research Ltd, Wallingford, January 1986.

RESEARCH

3   POWELL K.A. Shingle beach profiles under normal wave attack. SR Report, Hydraulics Research Ltd, Wallingford (In Press).
4   PRICE W.A., TOMLINSON K.W., and WILLIS D.H. Predicting changes in the plan shape of beaches, Proc. 13th Coastal Eng. Conf. Vancouver, 1972.
5   OZASA H and BRAMPTON A.H. Mathematical modelling of beaches backed by sea walls, Coastal Engineering, Vol 4, No. 1, 1980, pp 47-63.
6   BRAMPTON A.H. and MOTYKA J.M. Modelling the plan shape of shingle beaches, Lecture Notes in Coastal and Estuarine Studies, Vol 12, Offshore and Coastal Modelling, Springer-Verlag, Berlin, 1985, pp 219-234.
7   Hengistbury Head coast protection study, Report No EX 1460, Hydraulics Research Ltd, Wallingford, July 1986.
8   Effectiveness of groyne systems, Vol 2, Planning Guidelines, CIRIA/Sir William Halcrow and Partners, In Press.
9   PERLIN M and DEAN R.G. Prediction of beach planforms with littoral controls, Proc 16th Coastal Eng. Conf. Hamburg, 1978, pp 1818 - 1838.
10 ALLSOP N.W.H. Sea walls : A literature review, Report No. EX 1490, Hydraulics Research Ltd, Wallingford, September 1986.
11 OWEN M.W. The hydraulic design of seawall profiles, Proc. Symposium on Shoreline Protection, Inst. of Civ. Engrs. London, 1982, pp 129 - 136.
12 SMALLMAN J.V, ALLSOP N.W.H. and BRAMPTON A.H. The hydraulic design of offshore breakwaters in coast protection, Report No. IT 305, Hydraulics Research Ltd, September 1986.
13 POWELL K.A. and ALLSOP N.W.H. Low crest breakwaters, Hydraulic performance and stability, Report SR 57, Hydraulics Research Ltd, Wallingford. July 1985.
14 PIANC. Guidelines for the design and construction of flexible revetments incorporating geotextiles for inland waterways, Supplement to Bulletin No. 57, 1987.
15 ALLSOP N.W.H. et al (Editors) The use of rock in coastal structures, Summary of Seminar, Hydraulics Research Ltd, Wallingford, January 1986.
16 van der MEER, J.W., Rock slopes and gravel beaches under wave attack, Communication 396, Delft Hydraulics Laboratory, 1988.

9. Engineering conservation

A. G. ROBERTS, MICE, Chief Engineer, Canterbury City Council

SYNOPSIS. The implementation of coastal defence schemes can have a substantial impact upon the environment of the area to be protected. The paper contains a summary review of the current United Kingdom legislation in connection with coastal defence matters. The consultation process is considered, together with descriptions of the appropriate bodies both statutory and voluntary and their role in the protection of the environment. The paper explores in detail three case histories of coastal projects within the Canterbury area. The examples selected explore the range of environmental problems that the Engineer could encounter. The Author's conclusions are drawn from the roles of the Engineer and the Conservationist.

INTRODUCTION
1. As we move towards the 21st Century, we as a nation are becoming more and more aware of our heritage and our surroundings. All political parties are making pronouncements upon "green" matters and claiming conservation for their own, undoubtedly reflecting the views of the nation as a whole. The coastal defence Engineer is not isolated from such trends and has to reflect these changes in opinions and priorities. The work of the coastal Engineer can have a major impact upon flora, fauna, historic and archaeological remains and the geological form of the landscape.

THE LEGISLATION
2. In Great Britain the statutory obligations upon the Engineer are contained in the Wildlife and Countryside Act 1981 and the Town and Country Planning Act 1971. The Ministry of Agriculture, Fisheries and Food (MAFF), the Department of the Environment (DOE) and the Welsh Office have produced a guide entitled "Conservation Guidelines for Drainage Authorities" (ref. 1) which sets out in concise form advice upon conservation and wildlife matters for England and Wales. The aims and objectives which should be followed are summarised as follows:
  "[a] to further the conservation and enhancement of natural beauty;

[b] to further the conservation of wildlife and geological and physiographical features of special interest;
[c] to have regard to the desirability of:
   (i) protecting buildings and other objects of archaeological, architectural or historical interest, and
   (ii) preserving public rights of access to areas of mountains, moor, heath, down, cliff or foreshore and other places of natural beauty; and
[d] take into account the effects of any proposal on the preservation of rights of access and on the beauty or amenity of an area, on wildlife, features, buildings or objects."

In Scotland, the Scottish Office have similar objectives although no formal advice has been issued.

3. Interestingly the summary uses the phrase "to further" which implies a positive obligation. In considering these objectives, the Engineer has to obtain a valid statutory permission from the Planning Authority, MAFF, DOE or Scottish Office and if he is to minimise conflict has to act accordingly. Comprehensive consultation is therefore essential.

4. Particular areas requiring close attention by the Engineer are Sites of Special Scientific Interest (SSSI's), Environmentally Sensitive Areas (ESA's) and Areas of Archaeological Interest, which have come about under the Wildlife and Countryside Act 1981, Article 19 of the European Community Structures Regulation EC 797/85 and the Ancient Monuments and Archaeological Areas Act 1979.

THE CONSULTATION PROCESS

5. A proper consultation process is an extensive and sometimes lengthy act, involving a large number of bodies some of which may have aims and objectives which lead to conflict. The Engineer should attempt to find ways to avoid damage to the environment as well as considering how wildlife can be conserved and if possible enhanced by the incorporation of features for both habitat and wildlife. The following bodies provide a starting point for the consultation process.

6. The District Council, as the Planning Authority is the first contact and can provide valuable information upon local bodies and any environmental constraints that may be present within the site area.

7. The Nature Conservancy Council (N.C.C.), is able to give much valuable information upon all aspects of the natural environment, including any notable geology, flora and fauna and general wildlife matters. They will also be able to provide contact points with local voluntary bodies.

8. National Trusts and National Park organisations may have areas of interest at or near the works and should be consulted. These bodies are important in the context of the coastal process and care must be exercised by the Engineer to ensure as far as possible, that the direct or indirect effect

of the works on their areas is acceptable.

9. Heritage organisations such as English Heritage and Cadw may also have areas of interest at or near the works and the same consultations and care are necessary in the design process. Care has to be exercised by the Engineer to ensure that his works do not indirectly affect an adjacent area.

10. The Councils for the Protection of Rural England and Wales will also give advice on the impact of schemes on the countryside. Their comments should be sought at an early stage.

11. Local Nature Conservation Trusts together with the Royal Society for Nature Conservation form the largest voluntary organisations concerned with all aspects of wildlife conservation in the United Kingdom. Consultation with local trusts will provide valuable advice upon flora and fauna as well as ecological issues. They are often able to give more detailed information than that possessed by national bodies.

12. The Royal Society for the Protection of Birds can give much valuable advice upon the bird life of any coastal area, nesting sites and reserves and the facilities that are needed to ensure that bird life is not unduly affected. The Society can also provide advice upon local groups for contact purposes.

13. Before finalising any coastal scheme, the Engineer should discuss it with as many of the above organisations as is appropriate to ascertain the environmental objectives. Only then can the appropriate design be formulated. In this way the Engineer can try to merge the engineering objectives with those of design to maintain or improve the ecology of the area. It has to be admitted that some objectives can overrule others and compromise is necessary, either in engineering, or in ecological matters. In setting up a dialogue both the Engineer and Conservationist can understand the point of view of the other and move towards an optimum solution which is acceptable to both as being the correct one for the resolution of the problem. This dialogue is time consuming, but may avoid confrontation at a lengthy public inquiry or investigation.

SCHEME EXAMPLES

14. Canterbury City Council (C.C.C.) is responsible for some 18 km of coastline, situated on the North Kent Coast facing the North Sea. The area includes the coastal towns of Whitstable and Herne Bay. The topography of the coastline includes coastal flats, coastal clay slopes and sandstone cliffs. A main river has is outlet within the area and the coastal geology is varied. Being within the "affluent" south east of England, the coastline is under constant pressure from people who wish to spend their time within a coastal environment, with all the advantages and facilities that are normally associated with large towns. Canterbury itself has a strong tourist profile with about 2 million visitors annually, some of whom turn to the coast for part of their visit. The

coastline is therefore the subject of "people pressure". If one turns away from the normal visitor facilities and looks at the coastal ecology of the area, other facets become apparent. The following examples have been chosen to illustrate those facets and to describe the challenges that have faced the coastal Engineer.

15. <u>Eastcliff III</u>. At the eastern end of C.C.C's area, there is an outcrop of coastal sandstone cliffs known as Eastcliff III. The cliffs are about 1650 metres long and are comprised of up to 3 metres of Head material overlying London clay up to 10 metres thick, over Oldhaven, Woolwich and Thanet beds. Some 1300 metres of this length fronts agricultural land. At the eastern end of the cliff lie Reculver Towers, a site managed by English Heritage. At the western end lies the community of Bishopstone, an outlying residential area of Herne Bay. (Fig. 1 is an aerial view of this length.)

Fig. 1 - Aerial Photograph -
Herne Bay, East Cliff III, 1986.
Photograph by Fotoflite, Ashford, Kent, TN23 1ES

16. The coastal cliffs were largely unprotected from the action of the sea. As a result erosion occurred causing a landward regression of the coastal slopes. The eroded material provided a natural source of beach material for littoral transport in a westerly direction to other lengths of coastline where sea defences had been provided and beach

control systems using timber groynes existed. The land was in the ownership of C.C.C. and the hinterland was leased to adjoining farmers as grassland and for grazing. Generally this erosion of the coastal cliffs was of little consequence except at the western end, where if it were allowed to continue unchecked, the community of Bishopstone would, over the next 50 years, be lost to the sea. The people of Bishopstone were becoming more aware of the continuing erosion and the approach of the cliff top to properties. An increasing number of representations were made to C.C.C. for a sea wall and cliff stabilization scheme to prevent any further erosion.

17. The very act of erosion made the area interesting to the geologist and the site was one of the first to be declared as a S.S.S.I. by the N.C.C..

18. Discussions were held with the local area office of the N.C.C. to ascertain the ecological problems of the area. These discussions highlighted the strong geological interest, the interest of the ornithologists in the cliffs as nesting sites for birdlife and the dismay of the naturalist at the farming operations carried out by tenant farmers. Further discussions refined the problem to the conflicting views of the geologist and the home owner. Who was there first and which was more important to society then became the question. Taking into account that some 1300 metres of similar cliff were to remain and more importantly were not to be protected, persuaded the N.C.C. to agree to the construction of a short length of sea defences and the treatment of the overlying London clay and Head materials by re-grading and drainage. All the loose material was removed from the sandstone cliffs. In addition, C.C.C. ceased renting the cliff top for grass cutting and grazing and allowed it to revert to a natural state, to form a habitat for wildlife. The N.C.C. also agreed to attempt to foster more awareness of the local environment in the local community.

19. Fig. 2 shows the completed scheme which utilised a large rock graded rip rap revetment to protect the toe of the sandstone cliffs, and regrading and drainage of the London clay slopes, including the use of vertical Sandwick drains to drain the London clay unit downwards into the sandstone. The main contractor for the scheme was Edmund Nutall Ltd, with the specialist drainage sub-contracted to Cementation Ground Engineering. The scheme was essentially completed in 5 months at a total cost of £1.1 million. In order to dispose of the surplus excavated material from the regrading, permission was obtained from M.A.F.F. to deposit this material on the beach at the toe of the cliffs adjacent to the site, below high-water mark, so that it would be removed by normal sea action over the next winter period. Unfortunately, the next winter was mild and few winter storms occurred which has left some material on the foreshore. It is intended to ensure that this remaining small quantity is removed.

20. In engineering terms the scheme has been extremely successful in arresting erosion and stabilizing the overlying

materials. In ecological terms, not all of the objectives have been achieved although bird life seems to be returning to normal with the appearance of large numbers of nest sites within the sandstone cliff face. The cliff top habitat is also taking on a new shape and again the occurrence of nesting sites is encouraging. The N.C.C. as yet have been unable to commence an education programme within the community and it is hoped that the continuing discussions between both bodies will see some progress on that front.

Fig. 2 - Aerial Photograph - Herne Bay, East Cliff III, showing sea defences 1987. Photograph by Fotoflite

21. So far as the N.C.C. are concerned the loss of the 350 metres of eroding sandstone has not been compensated by the increased access that is now available. It was hoped that the slower weathering of the sandstone cliff face would still present new items of interest for further study. Unfortunately this has not been realised and the protected length the cliff face is starting to vegetate over, this is disappointing to both parties and further discussions are to be held to see if this trend can be reversed. The problem of the clay tip illustrates how easily understandings can fail and has led to a temporary lack of trust which has to be addressed.

22. <u>Swalecliffe</u>.   In contrast to the Bishopstone area of Herne Bay, the suburb of Swalecliffe is located further to the west, between the towns of Herne Bay and Whitstable and is on a coastal flat at a level just above that of highest

astronomic tide. It is typical of much 1920-30's development and consists of many bungalows and semi-detached homes. This form of development is attractive to the more elderly members of our society and the age profile of residents within the area is weighted to the elderly. The area has always been at risk to action by sea storms and suffered flooding in 1953, 1978 and 1979. Fig. 3 shows the general geographic form.

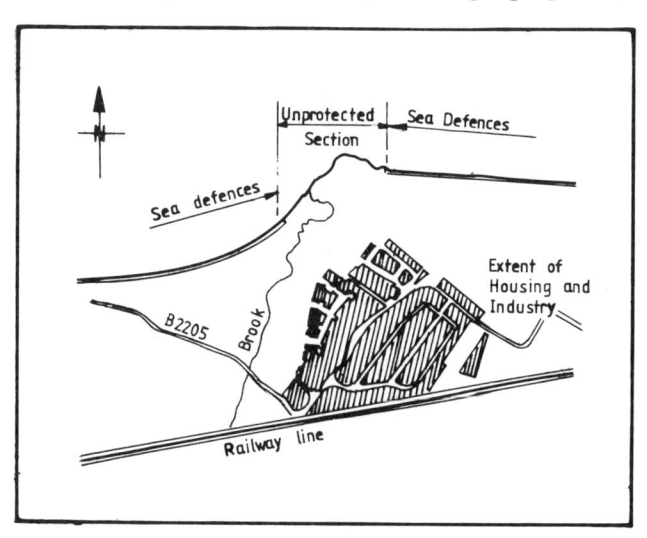

Fig. 3 - Swalecliffe, Whitstable, showing break in sea defences prior to 1986

It can be seen that the embankment of the Thanet to London railway forms a southern boundary to a shallow saucer shaped area containing much development, including a small industrial estate, containing most of Whitstable's industry. At first glance this description belies any possibility of environmental gain by the construction of sea defences. Clearly these increase the level of protection to the area and remove the risk of flooding by the sea. A closer examination of the problem identifies a narrow strip of land at the western boundary of the site, through which flows a main river, the Swalecliffe Brook. This river outfalls to the sea at Swalecliffe and at its mouth a small area of salt marshes exists with no sea defences. This gap in the sea defences was a source of concern to the residents, whereas the area was a source of interest to conservationists.

23. In preparing schemes to solve the problem, the initial proposal was to construct a sea wall across the estuary area, making provision for some form of sluice gate to control the flows within the river, as well as improving the existing defences. This solution was adopted as the Authorities method to resolve flooding problems. In civil engineering terms this presented no great problem and concrete walls, timber groynes

and the provision of a shingle beach to match the existing were proposed. This proposal found no favour with the Conservationists who objected most strongly, so strongly indeed as to persuade the Secretary of State to hold a public inquiry. The result of this inquiry was to reject the Council's proposals whilst recognising the need for some form of improved sea defences for the area.

24. In a positive response to the problem, discussions were held with interested parties including the N.C.C. These identified a two part solution, the first part being much as before in that the existing sea defences would be upgraded. The change came in the proposals for the area of the estuary where the suggestion of an objector was adopted in principle, albeit in a much modified form. This evolved into the closing of the gap between the two adjacent lengths of sea defences by the provision of a landward bund wall, constructed around the perimeter of the salt marsh. Not only did this proposal find favour with all, but it was also cheaper, despite some extensive landscaping works to the existing sea defences to reduce their harshness on the horizon from Swalecliffe itself.

25. Fig. 4 illustrates the defences to the area. The line of the bund can just be determined. The area of salt marsh is a haven for birdlife. Its ease of access means that local associations have in turn increased their study of the area. During the summer the area is a breeding ground for skylarks and meadow pipits which can be seen throughout the year. Finches frequent it, feeding on the seeds of the various wildflowers amongst the coarse grass, while pied-wagtails feed upon the insects which abound. These attract small predators like the kestrel, crows and magpies. Other small birds have been seen there swooping along the main river in search of small insects. In the winter reed buntings and other marshland birds can often be found.

26. Mostly the sea birds flock along the coastline and feed upon the mud flats or at the point of discharge across the shingle of the main river into the sea. However in bad weather the gulls can be found resting in the sheltered regions of the main river where it meanders through the salt marsh. The waders and the brent geese prefer to fly further afield onto the short wet grass land behind the nearby caravan site. As at many sea-side sites the gull family abounds. In the winter the bird numbers at Swalecliffe are swelled by many waders, including turnstones, red-shank , dunlin, curlew, oystercatchers, godwits, ringed plover and golden plover. A score or more of snow buntings have been observed over-wintering from their breeding ground in Scandinavia while great-crested grebe have been seen fishing off-shore with cormorants and the occasional red-throated diver. Whilst the construction works for the improvements to the sea defences were underway, a slavonian grebe wintered on the main river, rather disproving the previously expressed views of the Conservationists!

Fig. 4 - Aerial Photograph - Swalecliffe, Whitstable, showing sea defences. Photograph - Fotoflite.

27. It took a public inquiry to produce the necessary compromise between the Engineer and the Conservationist. The scheme adopted met all the engineering criteria, whilst the Conservationists compromised upon the "no change at all" stance. It is interesting to note that the site is now a wildlife sanctuary and is being considered for designation by the N.C.C. as an S.S.S.I..

28. Tankerton coastal slopes. The next example is at Tankerton, Whitstable where the coastal slopes extend for some 1850 metres mostly between 16 to 21m AOD dropping to about 6m AOD at either end. The average slope angle is between 13 and 19°. The length is composed only of London clay and since the early 1900's has been protected at the toe from wave action by conventional sea defences, which were largely constructed at the same time as substantial estate development to the hinterland at the top of the clay slopes. A road was also built as a part of this development some 30 metres landward of the slope top. Fig. 5 shows a general aerial view of the area.

29. Before protection, the slopes had been subjected to a cyclic process of erosion by sea action, consequent steepening and subsequent failure, enabling new material to be presented to the sea for the erosion process to repeat itself. The construction of the sea defences stopped the toe erosion by the sea, but as the slope gradient was greater than the angle of repose for London clay, failures could still occur on the face of the slopes. Stability was maintained by regrading the slopes to an acceptable angle and the provision of shallow

drainage systems on the face of the slopes. Some of the drainage was installed as early as 1929.

Fig. 5 - Tankerton Slopes, Whitstable, Kent
Photograph - Fotoflite

30. Most of the slopes were treated with regrading and drainage works, mainly in stages between 1962 and 1972. There are two areas left which have not been treated since 1962 - and these were in an unstable condition. C.C.C.'s Coastal Committee expressed concern and pressed for the preparation of a scheme to bring these unstable uneven areas up to the standard of the other areas, which having been stabilised and turfed, produced a popular grassed area for many visitors to enjoy.

31. In the early stages of the investigation, the links formed with the local societies revealed the existence of a nationally rare plant, Hogs Fennel (Peucedanum Officinale), which was abundant on parts of the unstabilized London clay slopes. The plant is a protected species and is illustrated in Fig. 6. Clearly the original requirements of C.C.C. were not going to be possible without the destruction of the Hogs Fennel. However, Wye College (University of London) was approached and now provide C.C.C. with all scientific advice upon the plant. The discussion process on the Tankerton Slopes was commenced with the N.C.C..

32. The conservation of this colony of plants is the joint aim of both Councils. As yet, no clear solution to the

problem has emerged. C.C.C. has embarked upon research into the characteristics of slope drainage systems, their efficiency and long terms characteristics and life. Wye College is looking at the biology of the plant including its propagation. What has become clear is that the plant appears to thrive best on a slightly unstable slope, where tension cracks allow the seed to take root in bare soil where the competition from other species of plant and grass is at a minimum. On the other hand, if the slopes were left to regress further landward, the stability of the road would become undermined, and pressure to carry out the works would grow at the expense of the plant colony. It is hoped that this research will guide all towards the best solution. Only by working together can the Engineer and Conservationist ensure the best compromise.

Fig. 6 - Peucedanum Officinale
(Hogs Fennel) At Tankerton Slopes

CONCLUSIONS

33. The foregoing illustrates how the attitude of the coastal Engineer has changed, reflecting the change in society's attitudes as a whole. Today, it is unthinkable for any proposal to proceed without all the environmental consequences being properly addressed by the Engineer. The best Engineer is now aware how environmental matters have to be considered and has responded to this new challenge. If one is critical of the new found relationship between the Engineer and the Conservationist, it is the latter who has yet to respond more positively to the needs of the Engineer. Like all good long term relationships, the need for give and take

has to exist upon both sides and in too many cases the Conservationist becomes entrenched in his aims. Sometimes compromise upon his part has to occur, if a solution is to be found to solve the problem. Only by working together can trust and understanding be established.

ACKNOWLEDGEMENTS
34. The Author would like to thank M.J. Bacon, MA, MPhil, MRTPI, City Technical Director, and S.E. Burnett, MSc, FGS, Research Assistant, for their help and encouragement given in the preparation of this paper.

REFERENCES
1. MINISTRY OF AGRICULTURE, FISHERIES AND FOOD, DEPARTMENT OF THE ENVIRONMENT, WELSH OFFICE. Conservation Guidelines for Drainage Authorities, p.5. HMSO, London, 1988.

10. Nature conservation in coastal management

K. L. DUFF, BSc, PhD, MIG, FGS, Assistant Chief Scientist (Earth and Marine Sciences), Nature Conservancy Council

SYNOPSIS. Britain's coastal areas are internationally renowned for their biological and geological features, but are very vulnerable to artificial changes brought about by engineering works. This paper looks at the effects of engineering structures and operations upon nature conservation, and suggests ways in which their effects could be significantly reduced.

INTRODUCTION
2. The coastal areas of Britain contain many features of nature conservation importance, ranging from geological features such as coastal cliffs to biological systems such as saltmarshes, dune systems, estuaries and intertidal areas. Many of these are of national or international significance in scientific and conservation terms, and present considerable problems of practical management, especially where development pressures and engineering proposals conflict with the safeguard of natural environmental systems. There are many instances around the coast where important wildlife habitats or geological outcrops have been destroyed or seriously damaged by construction works, and with the current growth of interest in environmental matters, such effects often give rise to intensely polarised reactions. This need not necessarily be so, since amelioration of effects if often possible, but it does require a firm commitment by environmentalists, engineers and developers to work together in attempts to achieve mutually acceptable schemes, as well as a sound understanding of the physical processes which operate. Joint efforts of this kind are still relatively infrequent in Britain, but offer considerable scope for improving the environmental acceptability of development schemes.

GEOLOGICAL FEATURES
2. Many coastal cliffs are important for geological studies, because of the rock sequences or fossil assemblages which they contain. Their height, lateral extent, and quality of exposure, which derives from continuing marine erosion, and ease of access, make such sites of the greatest

importance to geologists, for both training and research. Few comparable exposures exist elsewhere, with the possible exception of large working quarries which are frequently unavailable for study because of safety and access constraints. A considerable number of cliff exposures especially, on the southern and eastern coasts of Britain, are formally designated as nationally or internationally accepted standard reference localities (or 'type-sections'), which are used as standards against which to compare and calibrate rock sequences of comparable age throughout Britain, Europe or the World.

3. Practical conservation problems arise when cliff erosion rates are such that coast protection works are proposed. This occurs most frequently on the south and east coasts, where soft Mesozoic, Tertiary and Quaternary rocks are exposed. Traditional coast protection measures, such as mass concrete walls, revetments, cliff-grading, cliff-drainage, or groyne-fields, almost always leads to the immediate or rapid loss of any geological interest which may previously have occurred, as it is directly obscured or becomes hidden beneath a vegetation sward which develops on the graded and stabilised cliffs. Numerous important sites have disappeared over the past 50 years as a result of such works. This of course needs to be balanced against other considerations, such as threat to property, loss of agricultural land, and public safety, and such balances are the responsibilities of the Coast Protection Authorities and MAFF. However, since coast protection schemes are designed for 100% effectiveness, the question as to whether erosion rates can be reduced significantly, but whilst still allowing a small amount to continue in order to maintain cliff exposure, is rarely addressed. This is an area in which the Nature Conservancy Council (NCC) has been active in recent years, and where we have commissioned coastal scientists and engineers to investigate the practical feasibility of such a philosophy. We hope that coastal engineers will give serious consideration to the practicability of such an approach, and are keen to develop links between our staff and the engineering profession to enable us to investigate this further. Two initiatives are of relevance here.

4. The first application of such an approach was at West Runton on the North Norfolk coast, where an important section of Quaternary rocks was threatened by coast protection works in 1974. After lengthy discussions involving the NCC, the North Norfolk District Council, the consulting engineers (Mobbs and English) and DoE, design modifications were agreed for two lengths of the wooden revetment which was to be installed 10 metres seaward of the toe of the cliffs. These involved the reduction in numbers of facing planks in two lengths of the revetment, each about 150m long, in front of the two most critical parts of the geological section. The facing planks were reduced from 10 to 4, in the expectation

that this would permit increased water flow through the revetment, to allow the washing out of sediment that would otherwise accumulate at the toe of the cliffs behind the revetment, and lead to the build-up of a stabilised talus slope. The hypothesis was that the revetment would still reduce the wave energy very substantially, thereby slowing erosion, but would still allow gradual removal of fallen material.

5. The results of this work were monitored over a three year period from 1975, to assess the success of the modifications. The observations made, and the analysis of them, were reported by Clayton & Coventry (ref. 1), from which it is clear that definite conclusions are difficult to draw. However, it does seem clear that neither the normal 10-plank revetment nor the modified 4-plank revetment have been successful in halting cliff erosion. The cliff is still attacked at the base sufficiently frequently to remain steep and unvegetated, and the geological exposures remain visible and retain their scientific value. More interesting is the effect which the 10- and 4-plank revetments appear to have had on beach levels in front of the structure. Over the ten year period from 1976 to 1985, beach levels in front of the 10-plank revetments have fallen by an average of almost 1.1m, whilst only by an average of 0.75m in front of the 4-plank revetments. This appears to be statistically significant, and Clayton & Coventry concluded that the reduced efficiency of wave reflection of the 4-plank revetment causes less beach drawdown, and speculated that the standard 10-plank revetment may actually be too effective as a wave reflector.

6. In terms of developing a design which would enable specific sections of cliff within a protection scheme to erode at slightly higher rates than the remainder of the section, the West Runton work has not produced an immediate answer. However, it does highlight a novel approach which may repay study elsewhere, and also provides evidence that such modifications may not cause any deterioration in performance. We hope that engineers will give consideration to other innovative approaches to coastal protection design.

7. Since the results of Clayton & Coventry's work, the NCC have considered how best to pursue the issue, and concluded that it was necessary to commission consultants to investigate alternative approaches to coast protection in situations where SSSIs were potentially at risk. The objective is to provide sound information which will enable NCC to respond more effectively in its role as a statutory consultee over all coastal protection schemes which seek grant-aid from government.

8. Hydraulics Research Ltd have been commissioned to undertake this work, and consider all the alternative methods of coastal protection which may be employed in a wide range of situations, with a view to designing schemes which may be directly applicable in particular situations. The work will

also seek to identify novel approaches, and define areas where further research is required. As part of the project, Hydraulics Research will meet engineers from MAFF and from the Coast Protection Authorities, to explore options, discuss general issues, and generally raise the profile of alternative or low-cost approaches to coastal protection. The significance of sea level rise, and global climatic change, is likely to throw this general issue into sharper focus in coming years.

BIOLOGICAL FEATURES

9. The variety of biological features around the British coast is wide-ranging, and includes natural physical structures, such as sand dune systems, saltmarshes, estuaries, inter-tidal areas, softrock and hardrock cliffs, and major shingle structures. Each of these is important in its own right as a biological habitat, but is also important for providing the substrate on which a range of often vulnerable species of plants and animals subsist. All of these physical features are parts of a complex and dynamic system, in which four main components are involved - the landforms which exist, the processes which operate, natural changes in sea level, and sediment distribution and behaviour. All of these form part of an integrated system, and artificially-induced changes caused by human activity therefore give rise to effects which may be felt way outside the area directly affected by man. For good coastline management, all of these elements need to be properly understood, which generally requires that adequate monitoring data has been collected, preferably over a time-series. It is especially important that such studies, and engineering design parameters, take into account the whole natural coastal system within which a particular site is located, rather than undertake studies purely on the basis of local administrative or land-ownership boundaries. This point applies equally in the case of geological sites, where coast protection works can have much wider impacts through interruption of sediment supply and the process of terminal scour.

10. Within coastal biological habitats there are many types of development and engineering works which can affect them. This ranges from major projects such as tidal barrages, marinas and harbours, to smaller projects such as sea defence works, stabilisation of sand dune systems, reclamation of saltmarshes, or use of shingle structures for gravel extraction or the siting of power stations. Whilst a number of these may be fundamentally incompatible with the longterm existence of the pre-existing habitats, there are often ways in which the environmental impact of certain works can be ameliorated, and the establishment of closer links between engineers and conservationists is desirable. This should occur at an early stage in project planning, to ensure the possibility of maximum flexibility.

11. Because of the extent of coastal systems, the most effective way of developing better methods of coastal management is by moving towards more integrated forms of coastal zone management, in which local authorities, landowners, environmentalists and engineers are involved. As part of this, there needs to be an opening of minds, on all sides, to the possibility of more innovative methods and approaches, preferably via collaborative work and research, with the broad objective being to design engineering works which are more suited towards the needs of the environment. Conversely, management of fundamentally fluid geomorphogical systems, such as sand dune systems, should also be considered in a more original way. In the past, the tendency has been (even in the case of management by conservation bodies) to attempt the stabilisation of dune systems, and the minimisation of erosion, whether caused by human activity or by natural processes. In the medium to longterm, such management leads to fundamental changes in the form and biological interest of dunes, causing loss of pioneer habitats and species, and the spread of intermediate and late successional stages over much wider areas. The realisation that such changes were becoming increasingly widespread, and that knock-on effects were accelerating, has led coastal ecologists to conclude that a much more radical approach to the management of sand dune systems should be developed. Comparable perceptual changes are needed across the whole range of coastal zone management, amongst all those involved.

12. A good example of a more environmentally-sensitive approach to coastal conservation is the Hampshire Coastal Study (refs. 2-3), which considers a range of options for coastal defence and flood protection along the Hampshire coast. The study involved inputs from the County Council, Hydraulics Research, Portsmouth Polytechnic and NCC, and reflects the County Council's interest in the preservation of its coastline in as natural a state as possible. The work highlighted the significance of erosion along the coast, especially in the cliffed areas of Christchurch Bay and to a lesser degree in the East Solent and at Hayling Island. It also identified a major trend towards narrowing of the inter-tidal zone, by landward movement of the low water line, which is recognised on both shallow gradient coasts (ie saltmarshes) and on steeper coasts. Since these changes occurred at different times, and in different places, a single cause such as sea level rise is considered improbable. The most likely contributory factors were seen as dredging, construction of sea defences, growth and decline of saltmarsh plants, and sea level rise. The scale of visible change locally is enormous, the inter-tidal zone in some areas now only being a fifth of the width it was 100 years ago.

13. The study recognised that the division of the Hampshire coastline into various administrative districts did not correspond well with the natural divisions from a coastal

engineering viewpoint. It identified a number of "cells" along the coast, each being an area or stretch of coastline which experiences similar wave and tidal conditions, and which in general is only weakly affected by events in adjacent cells. Only rarely is such a cell wholly within one district council's boundaries and single cells often span several boundaries, including county and regional water authority limits; in such areas this may give rise to some difficulties in adopting a consistent coastal defence policy. This situation can be further exacerbated in areas, such as the Solent, where both private and commercial landowners shoulder the responsibility for coastal protection; similar problems arise in Portsmouth, Langstone and Chichester Harbours.

14. The value of shingle beaches as effective coastal defences was clearly recognised, and HRL recommended that wherever possible these beaches should be preserved and actively maintained by nourishment or recycling; both are methods of coastal engineering which are much more sensitive and 'sympathetic' to the safeguard of biological and geographical features than are 'traditional' approaches.

15. The continuing rate of erosion along the substantial lengths of coast still fronted by saltings was recognised to be a matter of particular concern. As the width of the inter-tidal zone decreases, and the saltings deteriorate, the sea defences along the landward edge of the marsh areas are subjected to increasing wave action. This in turn leads to the necessity to strengthen those defences, which is expensive, and which also causes major loss of biological interest within these fragile and valuable ecosystems. The HRL review advised that this problem is one which requires careful consideration, and that particular emphasis should be put on research to help understand the reasons for the erosion, and as low-cost schemes to re-establish a stable wide inter-tidal foreshore. Such schemes may need to be tailor-made so as to preserve or increase the biological value of these important areas. The HRL report also advised that before large areas are tackled, a variety of methods of restoring saltings should be attempted on a trial basis, and that these trials, together with the remainder of the Hampshire saltings, should be a vital part of the strategy for the future. The recognition of the need for engineering solutions on the Hampshire coast to be designed specifically with the environment in mind, and the ways in which the problem should be approached, is an important milestone, and one which should receive the support of all those working in environmental protection.

16. Coastal authorities in Hampshire are actively pursuing some of these new technologies, and many of their engineers have shown considerable skill and inventiveness in tackling local problems. This has been supplemented by improved lines of communication, at officer and member level within local

authorities, and the Standing Conference of local district councils (SCOPAC) which has been operating in this capacity in recent years makes a major contribution.

THE KEYHAVEN-PENNINGTON LAGOON SYSTEM
17. This case study demonstrates the fragility of certain coastal habitat types, and how careful design of engineering works can overcome major practical nature conservation difficulties.
18. Keyhaven and Pennington Marshes in Hampshire are managed as a nature reserve by the Hampshire and Isle of Wight Naturalists' Trust, and fall within the large Hurst Castle and Lymington River Estuary SSSI in the western part of the Solent. They cover 240 hectares of saltmarsh, mudflats, offshore shingle banks, and saline lagoons, and are important for their flora, breeding and overwintering bird populations, and saline lagoon plants and invertebrates. The saline lagoons are of particular importance, being a nationally rare habitat, and containing two rare species of invertebrates (the starlet anemone and the lagoon sand shrimp) which are so vulnerable that they receive special protection under Schedule 5 of the 1981 Wildlife and Countryside Act. The lagoon system is a series of connected ditches and ponds lying behind the seawall, and stretching for 1.5km. The ditches vary in width from 2 to 25 metres, and reach 40cm in depth; there are two connections through the seawall, which allow seawater to enter at high tide, and freshwater input is derived from ditches draining the surrounding marshland (ref. 4).
19. Saline lagoons are one of the most infrequent and most threatened habitats within Britain's coastal zone, being easily damaged by coastal protection or sea defence schemes, or by land reclamation schemes. Saline lagoons characteristically experience large variations in their water chemistry and other environmental parameters, and therefore contain specialised species. Six species of lagoonal invertebrate and one lagoonal plant receive special protection under the Wildlife and Countryside Act, and five of these seven occur in Hampshire, two at Keyhaven. The only way to ensure the safeguard of these statutorily-protected species is by retaining their habitats, and a series of management guidelines for the Hampshire saline lagoons is proposed for them, to be agreed by the NCC.
20. The low-lying coast between Keyhaven and Lymington is protected by an 8km reinforced earth embankment (the Pennington sea wall), which separates the low-lying marshes and agricultural land from the intertidal mudflats and saltings. The marshes have historically provided natural protection to the flood defences, but the steady erosion of intertidal marsh seen in many parts of Hampshire is now having serious effects at Keyhaven-Pennington, and the floodbanks have already been locally reinforced, mainly by

the placement of concrete-grouted stone blocks on their seaward face. Despite these measures the seawall is still vulnerable to breaching and overtopping, and the area behind the wall is vulnerable to flooding. A Southern Water Authority report released in 1986 states that the seawall is no longer an effective sea defence and, despite high expenditure on maintenance and repair, could fail under the conditions of a 1 in 5 year storm.

21. A six year plan for the reconstruction of the seawall, using various forms of protection including Enkamat, flexible concrete revetting, and a reinforced concrete retaining wall, was proposed in 1986, but would have led to unacceptable levels of impact upon the saline lagoons, and the NCC opposed the scheme as presented. Subsequently there have been lengthy discussions involving Southern Water, the County Council and NCC, over redesign of the scheme, and the possibility of joint funding; MAFF grant-aid is not available. The positive attitude being taken by all parties in these discussions are fundamental to the attainment of an environmentally sympathetic scheme, and are a model for work elsewhere.

22. Experimental works have now been emplaced to investigate ways of minimising disturbance to the existing lagoons, and also to investigate the colonisation potential of a newly-dug lagoon. The first problem has been addressed by the fixing of a tightly-woven nylon material against the seaward bank of the lagoon, which is here a channel 8 metres wide, supported by scaffolding poles. This is designed to prevent the movement of fine sediment particles from the seawall workings into the lagoon; the same technique has also been used across the width of the lagoon, in the hope that this will keep turbidity in the lagoon at a low level.

23. The plans for the new seawall are for it to be wider than the existing one, which means it is likely to encroach on the lagoon system, and this is the reason for the second experiment, to investigate colonisation of a newly dug lagoon. This has been created at the eastern end of the site, and is linked to the main lagoon by a narrow ditch. The number of organisms which colonise the new lagoon, the rate at which this occurs, and the density of population which results, are all being monitored regularly by staff from Southampton University and NCC.

24. The results of these investigations will determine the feasibility of strengthening the seawall without causing significant damage to the nationally rare habitats immediately adjacent to it, and will guide the decisions made on how to proceed. They do, however, indicate the advantage of initiating discussions over the design and installation of engineering works within sensitive areas at the earliest opportunity, and the growth of real co-operative action of this sort is to be encouraged. It must, however, be based upon a preparedness by all parties to be flexible, and will

gain nothing if such an exercise is merely an attempt to polarise or entrench positions. The progress being made at Keyhaven-Pennington should give encouragement to enter such discussions.

CONCLUSIONS
25. Over recent years, the incorporation of nature conservation considerations and requirements within engineering schemes has become increasingly frequent, and the examples cited in this paper show how this can work in practice. To continue this process of development requires commitment from both conservationists and engineers, mutual understanding of the objectives and constraints of each, and a willingness to be flexible, both in terms of philosophy of approach to problems and in an acceptance of the possibility of compromise. It is necessary to bear in mind, however, that compromise is not always possible, and that in spite of the best endeavours of both 'sides' it may not be possible to achieve agreement. Where that occurs, the strength of the relationship that has been forged will be measured by the capability of all parties to accept the situation, and still be prepared to try again in the future. For their part, professional nature conservationists are becoming increasingly involved in funding research into innovative and conservation-friendly methods of coastal engineering, and we look to the engineering profession to embrace the same philosophy. In this way, the increased public awareness of, and support for, environmental protection and wildlife conservation, is most likely to find practical application.

REFERENCES
1. CLAYTON, K.M. and COVENTRY, F. An assessment of the conservation-effectiveness of the modified coast protection works at West Runton SSSI, Norfolk. Nature Conservancy Council, Peterborough. 1986. CSD Report 675.
2. HYDRAULICS RESEARCH LIMITED. Review of the Hampshire Coastline. Volume 1 - Text, Figures and Plates. Hydraulics Research Limited, Wallingford. 1987. Report EX 1601.
3. HYDRAULICS RESEARCH LIMITED. Hampshire Coastal Review: Summary of Findings. Hydraulics Research Limited, Wallingford. 1987. Report EX 1627.
4. IRVING, R. The Hampshire Coast: a compendium of sites of nature conservation interest. Hampshire County Council, Winchester. 1987.

11. The Australian experience

I. R. W. MILLER, BArch, DipTP, ARIBA, FRAIA, MRTPI, MPAPI, AffilAMA, MEIA, Director, Planning and Environmental Services, Pak-Poy and Kneebone Consulting Group, Adelaide, and visiting consultant to Environmental Assessment Services Ltd, and P. A. MUMFORD, MSc, Consultant, Environmental Assessment Services Ltd

SYNOPSIS. This paper considers the lessons for UK coastal management that may be drawn from Australian environmental management experience and the use of environmental assessment as a planning tool for future coastal development.

INTRODUCTION

1. Every text book on environmental planning begins with an observation upon the speed with which environmental knowledge in the community has spread. The legislators, administrators and communities around the world are now apparently imbued with environmental wisdom. This is good, but probably old, news and is almost certainly inaccurate. Our contention is that we are in fact just beginning what will be a long process of learning about the environment and the processes of recognition and responsive management. This is particularly true for the coastal and marine environment.

2. A brief history of environmental management in Australia can be found in References (1), (2) and (3). This paper is a personal view about the future (and the present) of the environmental management task relating particularly to large scale, complex projects of which coastal zones offer many examples.

PROTECTION AND EXPLOITATION

3. Environmental protection law and the government agencies which have been established to administer it are but one aspect of environmental awareness. The authors of the Guide to Environmental Legislation (1) describe a conceptual framework of environmental control in which they perceive a "protective component" and an "exploitive component", see Fig. 1:

```
                Environmental planning
                   and protection


Disposition of                         Conservation of
natural resources                      natural and
                                       cultural resources



                    Development
Fig. 1.             legislation
```

4. The protective component may rely upon separate legislation and often separate administering authorities within each state and territory of Australia as well as quite distinctive acts and practices between the states for a host of protective functions including:-

. control over pollution of waterways
. clean air
. vegetation clearance
. noise control
. conservation of natural resources
. use of waterways
. conservation of cultural resources
. waste disposal
. control over hazardous and toxic substances

The exploitive component includes more rules for the disposition of natural resources and the facilitation of development, again by different agencies, within and between the states covering:-

- land use (other than for conservation)
- forestry
- building construction
- mining
- electricity generation
- fishing
- water resource utilization
- tourism
- regional economic development
- agriculture

Thus through the grouping of opposites a clearer understanding might be gained of the principles which bear upon the legal resolution of environmental conflicts. Some legislation which controls environmental processes includes elements both of planning (facilitation) and of regulation-examples being noise control, control of waterways, and planning control legislation.

5. The endless conflict between "development" and "conservation" is one of individual perception of where an acceptable balance should lie, bearing in mind that the two are usually competitive. The debate had shifted subtly to include within the meaning of conservation of resources the notion of renewal of resources as well. So far the only invocation to the renewal of resources in Australian environmental legislation seems to be an oblique one in enactments dealing with the conservation of native vegetation and of marine resources which, through being left relatively undisturbed, then have a chance of self-renewal through natural propagation processes.

PHILOSOPHIES OF DEVELOPED AND DEVELOPING NATIONS

6. Although there are obvious differences between the mechanisms of development and environmental control between the developing and the developed nations, it is interesting to ponder the underlying philosophical differences.

There appear to be three main philosophical ideas inherent in environmental management. They are the recognition that:-

(i). The exploitation of resources (whether renewable or not) upsets a complex system of natural and man-made dependencies and spreads ever wider but arguable progressively less intensively from the exploitive source.

(ii). The above can be anticipated and measured if the need is recognised and the will exists so to do. Environmental management depends upon systematic study and the application of effective means of control or stimulus in meeting agreed standards.

(iii). Environmental management is a socio-political issue rather than a strictly economic one. The choice between more or less environmental management is considered only when basic economic imperatives have been met.

These ideas help to explain the main differences which exist in recognition of the environmental consequences of major projects by developed and developing economies. Environmental policy development is gathering pace in developing countries and this is reflected in the recent insistence on behalf of the World Bank, Asian Development Bank and similar institutions that the projects for which they provide loans must be shown to be environmentally benign.

THE PROCESS OF PROJECT ASSESSMENT IN AUSTRALIA

7. Australia made the choice for committed environmental management in the early 1970´s. Australia has devolved, through federal powers inherent in the constitution, the responsibility for environmental screening of state government and private sector projects to the states and territories. The Commonwealth Department of the Arts, Sport, the Environment , Tourism and Territories (DASETT) administers federal environmental protection legislation through the Commonwealth Environment Protection (Impact of Proposals) Act, 1974 which is applicable to projects where commonwealth government decision-making is involved. These might include private projects, for example, requiring export licensing by commonwealth. DASETT issues environmental guidelines particular to each project, but in practice works through state and territory governments in liaising with private sector proponents, often using the state level authorities as executing agencies for the commonwealth.

8. All of the states and territories have agencies of government dedicated to environmental control, who administer acts of the state and territory parliaments for that purpose. Some have experimented with the combination of land use control, pollution control and environmental control within the single act and the single department. After experimentation and successive departmental and legislative review South Australia and New South Wales now both have consolidated legislation to implement planning consent and environmental authorization of those projects which are not dealt with at the local government level.

9. In New South Wales there is little formal integration of the planning and the pollution control legislation, although in practice a proponent must consider pollution control requirements when embarking on an environmental impact assessment because he will ultimately be called upon to apply for a license from the State Pollution Control Commission. Planning and pollution control are split between departments and ministers. By way of contrast South Australia's Department of Environment and Planning administers pollution control, heritage control, vegetation clearance control and national and state parks all within the one organization and under the single Planning Act. Further, since 1984 the Planning Act has integrated closely with the Clean Air Act. Western Australia, the Northern Territory and Tasmania have settled upon somewhat separate procedures for planning and environmental control.

10. In general terms (because the procedures do vary from state to state) environmental authorization is a staged process beginning with the issue of a notice of intent by government proponents and an application for a development consent or permit in the case of a private proponent. If a project is assessed by the relevant minister as requiring an environmental impact statement (EIS) in the case of environmentally sensitive projects - or a public environmental report of some kind for intermediate level projects - then the relevant planning authority is directed to make planning approval conditional upon the prior demonstration of environmental acceptability. The environmental authority is instructed to issue guidelines which set out the main environmental issues which are required to be addressed and the format expected for the ensuing environmental report.

When the environmental report is complete and acceptable to both the proponent and the authority in principle, it is placed on public exhibition for up to three months and advertised in the press.

11. Public submissions and departmental submissions are invited and responded to by the proponent in a new report which is sometimes called a Supplement. (In New South Wales a Supplement is not called for because aggrieved parties have greater opportunity to litigate). The environmental authority then assesses the proponent's responses to submissions and prepares an internal report to its minister advising whether the draft environmental impact statement together with the Supplement meet with the department's requirements in addressing satisfactorily the relevant environmental issues. Having considered the draft EIS, the submissions, the response to submissions and his department's own report, the minister determines what amendments need to be made to the EIS and, if satisfied, then signifies to the proponent in writing that the statement is officially recognised. Only after this can a planning consent be considered and granted. NSW uses a system of "designated development" - ie projects which are specified in the Environmental Planning and Assessment Act as requiring an EIS. There is no ministerial discretion.

SOME CONCERNS ABOUT THE AUSTRALIAN ENVIRONMENTAL EVALUATION PROCESS

12. The process of environmental clearance in Australia is subject to concerns which appear to find some common ground throughout the states and territories. The main concerns can be summarised as:

a) Excessive duration of the environmental authorization procedure. Project clearance can take in excess of two years from commencement of the EIS procedure.

b) Uncertainty in the end result, particularly in South Australia. For projects requiring an environmental impact report of some kind environmental clearance is a pre-requisite to the issue of a planning consent (Crown projects excepted). A project whose environmental impact statement has received official recognition can still meet with a planning refusal and therefore fail. Furthermore, one which has received environmental clearance and planning consent is still subject to appeal be representors (third parties) on planning and/or environmental grounds. In NSW an EIS does not receive official recognition, but there is greater opportunity for litigation.

c) The various environmental acts outline the key outputs required, but are silent about the process or steps for getting there, although the commonwealth legislation is more specific. Whilst the absence, generally, of detailed procedures and regulations imbues a flexibility which in itself is desirable, there is concern among industrialists and conservation groups that the absence of clear guidance as to procedure leads to uncertainty and can be costly in the waste of time and effort in the early stages. For example criteria indicating the sorts of projects which may be subject to formal environmental clearance are not generally specified in legislation.

d) Concern has been expressed that the public consultation procedures in common practice provide insufficient opportunity for the public to be involved in the decision upon the level of assessment appropriate to each project and that follow up of public submissions is uncertain and not thorough. At the root of public concern is the realization that the process of environmental impact assessment involves a continual stream of decisions by the government in consultation with the proponent and the proponents's consultants, to which the general public is not privy.

e) The EIS procedures are too demanding for those simple projects which, nevertheless, require some form of environmental analysis and assurance involving public consultation. An intermediate-level procedure involving less study, cost, documentation and uncertainty is called for.

f) Not all states have adopted procedures for ongoing environmental monitoring of projects which have continued substantive environmental impacts. For example the monitoring of hydraulic processes in an estuary after works which alter the tidal regime have been constructed would appear to be a reasonable requirement in order that changes in erosion and deposition patterns can be compared to those predicted and unforseen changes can be observed in time to allow for any remedial action required. Similarly the discharge of benign effluents into estuaries and coastal waters should be continually monitored to ensure acceptable dispersion and dilution rates.

g) In cases where an environmental assessment entails the comparison in environmental terms of two or more technical options, the environmental appraisal can shift subtly from a decision about whether the project meets acceptable standards of environmental impact to one which measures the comparative environmental implications of each option resulting in the inevitable preference for that demonstrating least environmental impact. There is an important philosophical difference between an environmentally superior project and an environmentally acceptable one. It is arguably not the business of the environmental authorities to insist on the superior alternative (which may be vastly more expensive than the other options).

h) The situation with regard to Crown development is not clear. There is no uniform provision throughout Australia for public comment to be sought on Crown projects unless an EIS is called for. As in Britain, the Crown enjoys various immunities from planning and environmental controls.

i) There is a perception that EIS's are promotional documents rather then scientifically objective assessments. The fact that the proponent prepares or pays the cost of preparation of the EIS often results in a glossy report which can mislead as to its true purpose. The EIS prepared by a proponent or his agent is sometimes accused of "whitewashing" the project.

j) Concern has been expressed that the indenture agreement process used increasingly between governments and project developers can lead to circumvention of the EIS process because the government maintains an interest in the projects's success, and the end result is more certain.

THOUGHTS ABOUT THE FUTURE

13. Several of the state governments are actively reviewing their environmental control and monitoring procedures and doubtless successive state and federal governments will apply their own different perspectives to the task of providing the most appropriate systems for their states and their times. South Australia's review of the environmental evaluation procedure is a good example (7).

14. The authors' view is that environmental planning in Australia may have become over-institutionalised, perhaps through being over-politicised. The degree of environmental planning procedure not being commensurate with the size or probable impact of the project.

15. Below are some thoughts on how the environmental evaluation of deserving projects might evolve in the future:

(i). An enduring difficulty is the manner in which the public (affected or just interested) are brought into the environmental assessment process. Arguably the public should have a say in which projects should be subject to a form of formal environmental impact study, as it is the public who will be affected by the project in the end. Equally the public must be given an opportunity to appraise the environmental assessment and help shape the agreement ultimately reached between the authorities and the proponent. The process of public consultation must be made a more evolutionary and continual one, rather than relying on a single public exhibition where those interested are called upon to articulate their full concerns formally, once, in writing. The flow of information in this process in one-way.

(ii). An example of how public consultation is being organised in a beneficial way in Australia and certainly in Britain, albeit at the local residential level, is seen in the approach taken by road construction authorities to local area improvement projects. The affected public's views are sought early in an organised way to street closures, traffic management measures, landscaping plans and so on, and early consensus reached on the most acceptable solutions.

(iii). Little of the enormous bank of data, knowledge and experience accumulated by the administrators seems to be made available as a matter of course to the proponent who is expected to gather afresh much of what has been gathered and sifted through before, and reproduce responses to minor issues for the same assessors to respond to. NSW has now introduced a system for EIS's related to coal mining projects, where archival information is made available to assist proponents.

(iv). The procedure of "scoping" followed in the United States, where a project description is circulated early to all interested agencies who are asked to identify only their major concerns, seems to have some merit. Routine and trivial environmental and social impacts are ignored or given little weight. The process of scoping leads to the production of environmental guidelines which, by consensus among the authorities, includes the most important environmental concerns related to the project in terms of design modifications and environmental investigation. To some extent this process is being carried out in Britain as in, for example, the "affected authorities" consultation exercises recently carried out as part of the environmental assessments for certain major coastal development projects which are the subject of private parliamentary bills.

(v). There is no justification or support in Australian planning and environmental law for authorities requiring, from an array of options, the environmentally superior option when they and their legislation fail to set clear standards of environmental acceptability. If a proponent were faced with a choice of three sites for a yacht harbour/tourism project and was required to prepare an EIS through which a choice of site on environmental grounds would be assessed and if the assessing authority were to apply clear criteria of acceptability, then (in theory) all three locations might be demonstrated to be acceptable in environmental terms. The choice would then be the proponent's to make on a combination of environmental, technical and cost factors (the last two factors being strictly beyond the environmental assessor's concern). In practice a choice by the proponent is usually pre-empted by the environmental assessment procedure.

(vi). A similar argument is suggested when examining the inclusion in most Australian environmental guidelines of the requirement for statements of project substantiation. Substantiation, being the demonstration of need or demand is in any case somewhat apart from the investigation of environmental effects and the two should not be confused. Nowhere in planning codes is it mandatory for an applicant to satisfy a planning authority as to demand for particular facility proposed; merely that its existence will not prejudice orderly, proper and economic development for the community. Substantiation goes well beyond the reasonable demonstration of orderly and economic consequences. Courts and tribunals of appeal, although frequently beset with arguments of need, are rarely persuaded by such argument, preferring to judge upon planning merit and compliance with the law, as they are directed to do by the relevant planning legislation and by their own charters. Demand is subjective, changeable over time, and reliant upon entrepreneurial and management effort. EIS clearance procedures should not concern themselves with substantiation, beyond a reasonable assessment of the socio-economic impacts of the project not proceeding, which the relevant authorities are rarely qualified to judge.

(vii). The uncertainty of the whole Australian environmental clearance procedure is a central concern. Perhaps what is required is a shift in philosophy from "prove to me that your project will be environmentally acceptable" to "let us plan together to make your project acceptable in environmental and planning terms". This latter philosophy opens the way to the environmental clearance process becoming a progressive sequence of clearances so that potentially unacceptable projects are identified and sieved out earlier, and potentially acceptable ones speeded on their way.

(viii). The realization is growing that environmental clearance must be followed by an effective compliance procedure lasting the life of the project. Compliance, for example with receiving water standards or with design specifications, can only be guaranteed on behalf of the affected community if there is a requirement upon the proponent for a continual monitoring programme involving (as necessary) measurement, amelioration and practice and reporting. The authority's responsibility would be for verification, followed by discontinuance and enforcement procedures where required. Environmental, and indeed coastal, management (like personal hygiene) must be regarded as a continual management process rather than a once-off examination and certification.

(ix). The fear that proponent's environmental reports tend to "whitewash" projects because they are not sufficiently independent is largely unfounded as the Australian environmental clearance process tend to neutralise any such bias. The proponent remains the practical agency to commission the environmental assessment. First, there is no agency other than the proponent (or his professional advisers) who can describe the project and the proponent's intend faithfully. Further, the procedure of environmental investigation does enable, through cooperation between the environmental investigators and the project designers, project modification and improvement to take place continually throughout the study and project design period. This should result in an improved project which avoids or at least mitigates many environmental impacts which, if it were not for the study process, would not have been recognised.

Second, there is no agency other than the proponent who could be put upon to pay the sometimes extremely high study costs. Finally, and most importantly, the arbiters of the degree of balance and objectivity which has been attained in an environmental report remain the community, the government departments who are consulted as a matter of course, and the minister. This, provides the ultimate safeguard against self-interest and bias in reporting. If the community, the bureaucrats and the minister do not like the report they receive they, collectively, can reject it.

RELEVANCE OF AUSTRALIAN ENVIRONMENTAL MANAGEMENT EXPERIENCE TO UK COASTAL MANAGEMENT

The environmental objectives

16. The environmental objectives of coastal management are common between Australia and the UK:

- Conservation of the coastal ecosystem.

- Preservation (not quite the same as (i) above).

- Improvement of the existing environment.

- Social amenity.

However, the emphasis in UK will have a slight difference to that in Australia.

17. Conservation

(i). Little of the coastal ecosystem around the UK can be said to be unaffected by man's activities, although the UK coastline remains rich in fauna and flora specifically adapted to a marginal existence. The recent growth in the designation of areas of estuary and open coastline as Sites of Special Scientific Interest (SSSIs) illustrates the importance of these areas for primary conservation.

(ii). Much of the open Australian coastline comprises large embayments between headlands which have a discrete dynamic stability within the compartment. Sediment only passing out of one compartment and into another during extreme storm events. The dominance of short period wave driven littoral drift on most of the British coast is not such a feature on Australian ocean beaches. On most Australian beaches which are open to the Pacific, Indian or Antarctic oceans, the longer period ocean swell waves dominate. Exceptions being beaches such as those at Adelaide in Gulf St. Vincent, South Australia, where coastal processes are very similar to those found around much of the UK coast.

(iii). Australian debates about causes of coastal erosion have concentrated on loss of sediment input from rivers, wind erosion of de-vegetated dunes and possible exhausting of sediment reserves in submerged Pleistocene beaches offshore. In the UK concern about coastal erosion has centered on interruption to littoral drift and loss of sediment supply from an updrift eroding coastline.

(iv). One of the primary coast protection options for relatively open, undeveloped Australian beaches has been the re-vegetation of sand dunes previously damaged by human activity. This is sometimes partially "improvement" as well as conservation. The re-vegetation first seeding adopted is Marram grass, <u>Ammophila arenaria</u>, which is not native to Australia and is a more effective sediment trap than the local spinifex. It can be observed from comparison with late 1930s aerial photography of (then) remote beaches that recently re-vegetated dunes are often larger and support vegetation further seaward than the original.

(v). Few areas of UK coast allow dune re-vegetation, or similar conservation techniques to be the primary coast protection strategy. One area of UK coastal management where the debate still centres on geomorphological conservation is where the Nature Conservancy Council are anxious to maintain the "living geology" of certain cliff faces and other coastal features by allowing coastal erosion to continue un-altered.

(vi). The main "conservation" drive on the UK coast concerns coastal flora and fauna under threat from marine pollution and loss of estuarine/inter-tidal habitat.

18. Preservation

(i). Preservation of an already altered ecosystem is the most familiar UK coastal management strategy. The importance of preservation of an existing stable and natural ecology cannot be overstated. The destruction or adverse modification of an existing habitat, even if partially man-made (analagous with urban wildlife), should not be disregarded. Hoping that displaced species will sort themselves out is likely to be an inadequate approach. Similarly, proposals which apparently benefit one species must be considered in the light of the impact of the stimulated species on those not directly benefiting.

(ii). The UK coastal manager must devise strategies for preserving coastal development originally built without regard for natural beach fluctuations, existing beaches held in place by earlier coast protection works and stretches of coastline adversely affected by others´ updrift works. To some extent Australia has managed to avoid many of these constraints.

(iii). The common instruments of preservation (groynes to maintain existing beach levels and sea walls to preserve an existing shoreline) once embarked upon, tend to commit the manager´s successors to the long term maintenance of the system.

19. Improvement

(i). Improvement of the existing coastal environment is an option considered with increasingly frequency these days, and one for which there is considerable scope around the UK coastline.

(ii). The coastal manager's instruments of improvement can be beach nourishment (where beaches are re-built to historic levels or above using similar or coarser material than the original littoral sediment), introducing sand by-passing of existing artificial obstructions to littoral drift (harbour piers etc) and the seasonal updrift return of beach material within a littoral compartment.

(iii). Other improvements in the coastal environment may be restoration of existing, or creation of new, habitats for coastal flora and fauna, creation of nature reserves, etc. However, any such improvements must take into account possible resulting changes in the existing balance of species.

(iv). Another area offering considerable scope for improvement on the UK coastline is existing coastal development. Whilst many UK coastal settlements are rightly famous for their outstandingly attractive character, some of Britain's coastline is blighted by developments ranging from the merely inappropriate to the unredeemably awful. This may impinge more on the role of the planning authority rather than the coastal manager.

20. Social amenity

(i). There is a need to consider human activity as an integral part of the coastal ecosystem. Grants for coast protection works under the terms of the Coast Protection Act 1949 were considered only for work strictly necessary for coast protection and not for items of public amenity. This has encouraged a utilitarian approach to coast protection works, even on the most scenic frontages. Fortunately, most strictly coast protection measures which can be compatible with public access and recreation. It would be better to consider the social amenity of the coast as an integral part of the overall coastal management strategy. This appears to play a more central part in Australian coastal management.

THE ROLE OF ENVIRONMENTAL ASSESSMENT IN UK COASTAL MANAGEMENT

21. In planning to achieve the environmental objectives it is necessary to have a mechanism to consider the effects of any proposed works or development on the broad environment. General rules alone are insufficient, as the nature of coastal works are very site specific. Further, some method is needed to define the existing environment and identify the sensitive aspects in the local ecosystem when considering a specific development proposal. The method chosen in Australia, the USA and now in Europe is environmental assessment (9). (the word "impact" having been deleted from the official European description).

22. The object of environmental assessment is to provide the opportunity to look broadly at proposed project and consider what the effects of the project are likely to be on the whole range of environmental aspects, and to give an opportunity to amend the proposed project at the planning stage to mitigate adverse effects. EA gives the framework to ensure that the full spectrum of possible effects are considered. Ideally, an EA needs to be carried out by a body reasonably independent of the proponent's internal organization in two phases (see Fig 2.):

(i). A preliminary assessment on the conceptual scheme, to ensure that an environmentally realistic project emerges and to identify the areas of particular sensitivity.

(ii) A full assessment of the detailed scheme for any consent applications and public scrutiny.

23. Australia's environment and planning policy which has included formal environmental impact assessment for some 10 years enables retrospective analysis of its successes and failures (8). It is advantageous to be able to consider the Australian experience now that formal environmental assessment has been introduced into the UK development planning process.

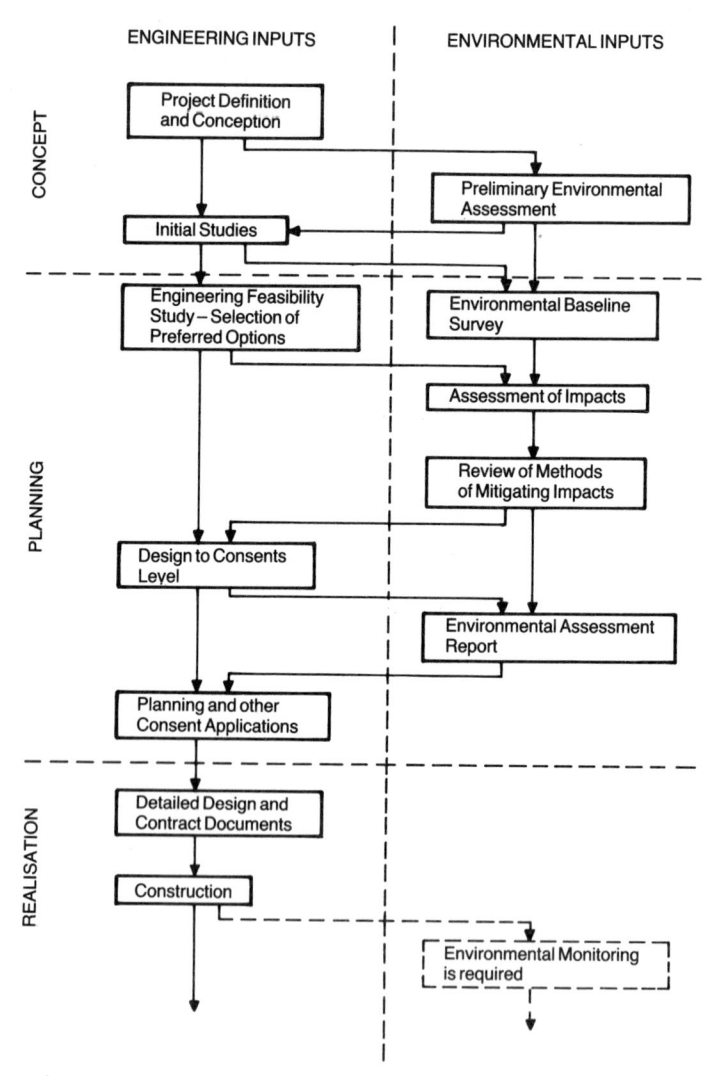

Fig. 2.

24. Since 15th July 1988, EC directive 85/337 (and its implementing legislation) has made formal environmental assessment mandatory for certain major development projects in UK. Coastal projects, except for major port works for vessels of over 1350 Tonnes, are not specifically included amongst those listed for mandatory EA in Annex 1 of the directive. However, almost all likely coastal development works can be classified under the categories in Annex 11 for projects requiring environmental assessment when the probable effects on the environment may be significant.

25. The requirements of EC directive 85/337 are simpler and more flexible than those of the equivalent Australian regulations. Similarly, fewer project types require mandatory assessment under the EC directive than under the Australian system.

SUMMARY

26. The authors believes that the procedure commonly in use throughout Australia for environmental vetting is likely to change to become less formalised, less orientated to a final reporting stage upon which a single yes or no decision rests, and more a progressive process providing early indications of concerns and important issues during which the community, the authorities and the proponent will interact. The routine canvassing of relatively minor issues should diminish. More emphasis is likely to be placed upon the setting of measurable criteria of acceptability and, when there are major options to be chosen between, environmental acceptability will supplant environmental superiority. After environmental clearance has been issued the environmental management plan should impose continual and long-term monitoring responsibilities on the proponent in the achievement of prescribed standards, and there is a case for stronger discontinuance and enforcement procedures. Environmental clearance and planning consent should become a single procedure, as will environmental and planning enforcement. Institutional strengthening should assist the interchange of information and understanding rather than ever more onerous procedure. The cause of legal uniformity in the environmental and planning management area is anticipated to progress across the states and territories and between the states and the commonwealth.

27. The flexibility of the environmental and planning procedures inherent in the implementation of EC directive 85/337 should allow the effective use of environmental assessment as a coastal management tool whilst avoiding some of the problems revealed by the Australian experience.

REFERENCES

1. Guide to Environmental Legislation and Administrative Arrangements in Australia. Second Edition. Australian Environmental Council. Report No 18. Aust Govt Publishing Service, Canberra 1986.

2. GILPIN A. Environmental Policy in Australia. University of Queensland Press 1980

3. PORTER C. F. Environmental Impact assessment. A Practical Guide. University of Queensland Press 1985

4. Environmental Planning and Management and the Project Cycle. ADB Environmental Paper No 1 Asian Development Bank. Sept 1987

5. Environmental Planning and Management. Regional Symposium of Environmental and Natural Resources Planning . Asian Development Bank. Feb 1986

6. JAKEMAN A. J., PARKER P. K., FORMBY J. & DAY D. Resource Development and Environmental Issues. Opportunities and Constraints in the Hunter Region NSW. ANU 1987

7. Environmental Impact Assessment in South Australia. Report to the Minister for Environment and Planning. Review Committee, August 1986

8. Environmental Legislation and its Impact on Management. Proc. Seminar, Institution of Engineers, Australia, Sydney, October 1987.

9. MCKEMEY M. D. & MUMFORD P. A. Environmental assessment - a basic guide. Municipal Engineer. Vol 5. Feb 1988.

12. The deterioration of a coastline

M. J. WAKELIN, BScTech, FICE, FIWEM, Chief Engineer, NRA Unit, Anglian Water

SYNOPSIS. The coastline of Anglia runs from the Humber to the Thames and a large proportion of it lies below high tide level. It presents a constantly shifting pattern wrought by natural forces. Its evolution is affected (albeit only slightly) by the works of man. Conversely, the changes in coastline, past and future, have had and will have a profound effect on the social, economic and recreational activities of man. The paper describes these changes, some of which are understood.

THE NEED FOR UNDERSTANDING

1. As people, we take a selfish view of our coastline and hinterland. We ask ourselves "what can they do for us?". They do, of course, provide us with space in which to live and work: but at a price. The low-lying parts (within Anglia, some 1600 km including estuaries: see Fig.1, also ref. 1.) must be protected against tidal flooding if we are to continue our habitation of them. The effects of failure to meet this need were painfully felt in 1953 and again, albeit to a lesser extent in 1978 (Figs. 2 and 3). Where erosion of the coastline threatens our way of life we must decide what action, if any, we wish to take for the preservation of land and property.

2. Many millions of pounds have been invested in works of sea defences and coast protection; many more will be needed in years to come. There is a need to spend £150 million in Anglia over the next 5 years. To optimise what we spend in effort, time and money, it is first necessary to understand the mechanisms at work.

DETERIORATION

3. The choice of this word in the title is perhaps an unfortunate one. From a detached point of view, the changes which take place are just changes; neither good nor bad. Nevertheless, a public authority, charged to provide a public service, must, of necessity, take a less impartial view. Those changes which are inimical to human activity must be regarded as undesirable.

COASTAL STUDIES

Fig. 1. Extent of coastal works along Anglian coast line

4. It is by no means true that the whole coastline is "deteriorating". Significant frontages are accreting and here there is no need for human interference. However, the coastal engineer cannot afford the luxury of "averaging out" and balancing an accreting frontage on the one hand with an eroding one on the other. The needs of all frontages must be considered individually and, where appropriate, action must be taken to oppose the natural course of events. In deciding what action to take, due regard must be paid to worthwhileness; it may be desirable to do nothing. Indeed, future decades may see an increasing use of the "zero option".

5. Deterioration, if we must use the word, affects both the natural coastline and the various artifacts with which man has presumed to tinker in the scheme of things. Let us consider first, the effects of change on these coastal structures.

EMBANKMENTS

6. Of all those frontages protected against tidal flooding, earth embankments, in one guise or another, form the defence in the overwhelming majority of cases. Out of 1600 km of sea defences, 1400 km are constructed in the form of embankments (ref.1.). This is not surprising since the first reaction of one wishing to control flooding (be he a Roman or an inhabitant of the 20th century; a child playing on the beach or a professional civil engineer) is to construct an embankment with whatever material is lying close at hand. Those who have had experience of this activity (to protect a sand castle or for some other purpose) know that such structures have finite lives, even with good maintenance.

7. At one extreme are silt embankments such as those along the West and South coasts of The Wash. These were constructed to a shallow slope, are often grazed, are not fully exposed to the North Sea and (of greatest importance) are protected on the seaward side by a high out-marsh. The lives of these defences may well be measured in hundreds of years.

8. At the other end of the scale (sandcastles apart), large areas of Essex are protected by clay embankments with steep slopes. The upper parts of these embankments dry out in a hot summer and exhibit fissures of alarming proportions (e.g. 80-100mm wide and 1.5-2.0m deep). The life of such a structure may be as little as 20-25 years.

9. Embankments which contain organic material and, even worse, are founded on unstable strata, can deform badly over only a few years. Indeed, where the foundation is bad enough, there is a limit to the height of embankment which can be built since any added height will be negated by an equal settlement of the underlying strata. This

COASTAL STUDIES

Fig. 2. Consequences of failure (i) Kings Lynn Parish Church, January 1978

Fig. 3. Consequences of failure (ii) Heacham caravan park, January 1978

description fits the tidal embankments surrounding Breydon Water in Broadland and the problem was encountered in the design of Thames Tidal Defences along the Essex marshes.

REVETMENT

10. Of the earth embankments described above, about half are protected on the seaward face (and some on the landward face too) by a form of revetment. The purpose of this section is not to describe the different forms of revetment (ref.2); rather it is to illustrate the way in which the standard of such protection can decline.

11. The lifetime of revetment will depend largely on the effort put into maintaining it. Provided the design is right and external conditions (such as beach levels) do not change, revetment which is properly maintained should last for many decades. However, maintenance is usually labour-intensive and therefore expensive. It is the author's view that, in general, maintenance standards of revetment have been allowed to decline.

12. Along the Thames frontage, damage caused by the October 1987 hurricane revealed deterioration over extensive areas of concrete block revetment. On that occasion, emergency repairs cost £170,000 but to put right shortcomings elsewhere will cost many times that. A survey is now in hand.

13. Humber bank revetment, originally constructed from Scunthorpe blast furnace slag, requires constant attention. Again, the work is difficult and expensive.

14. The maintenance of limestone revetment along those tidal rivers which discharge into The Wash (i.e. Great Ouse, Nene, Welland and Witham) is not quite so labour-intensive but does deteriorate rapidly without proper maintenance. All these rivers are bounded by flood embankments which are themselves protected by stonework. To seaward of each of the outfalls of these rivers the stonework is extended in the form of training walls. Their purpose is to contain the natural tendency of these rivers to meander in The Wash raising river levels and hindering navigation. Here is an example of a quite different form of "deterioration" successfully controlled.

HARD DEFENCES

15. Hard defences are, by definition, inflexible and therefore not the sort of structure one would choose to insert into the shifting environment of a coastline. They should be seen as a last resort, to be used only where a "soft option" is not available. We are asking for trouble by putting an unnatural obstacle in nature's path and, of course, we get it.

16. Some 6% (i.e. 104 km) of Anglia's sea defences are "hard" and all have a relatively short life-expectancy. Those which were built in the aftermath of the 1953 disaster have now had the best part of their lives.

Indeed, many have already been replaced; usually by a new hard defence built on the seaward side of the remains of the old one. It is significant that wherever this has been done, it has also been necessary to reconstruct the seaward toe at a lower level in the beach.

17. Deterioration takes various forms and it is instructive to observe which particular agent of change ultimately causes the demise of a structure.

18. Concrete deteriorates by abrasion and evidence of it can be seen at all arrises and joints. This is particularly noticeable at lower levels exposed to more tides and where the concrete has been cast in situ (presumably under severe weather conditions, between tides and with consequently poor quality control). However, this abrasion is usually only dangerous where it leads to the breakdown of a seal between ajoining sections which, in turn, allows the leaching of underlying material and the build-up of dynamic wave forces under the concrete.

19. This sequence of events can easily combine with the effects of a lowered foreshore to allow the undermining of stepwork and the creation of terrifying cavities. This occurred at Huttoft near Sutton-on-Sea in 1982 and again at Mablethorpe in 1984. On both occasions, a major breach was averted only by timely repair work.

20. Steel sheet piling deteriorates by corrosion and abrasion. At Chapel Point, a £1M project was necessitated, at least in part, by an exposed steel sheet pile wall corroding almost right throught. Abrasion by beach shingle was the partial cause of failue at Felixstowe where both steel sheet piling groynes and toe-piling were simply worn away.

21. Corrosion of concrete reinforcement is sometimes all too obvious a manifestation of structural decay. The phenomenon is ubiquitous and is not confined to coastal defence works.

22. Hard defences of all kinds carry within themselves the seeds of their own destruction. The greater reflectivity of the seaward face causes turbulence under wave action and encourages erosion of beach material. A vicious circle is created when that erosion causes the exposure of toe piling which presents a vertical face to the incoming waves which, in turn, exacerbates the turbulence ...

23. Where it has been necessary to create a hard defence, the best way to enhance its life-expectancy is to cover it with sand; thus deceiving the waves into the belief that the defence is soft. An obvious solution to the problem but, sadly, an expensive one. As our knowledge of coastal processes advances, it is to be hoped that we shall devise cheaper ways to encourage beach growth thus extending the lives of existing defences and reducing the need to build new ones.

GROYNES

24. The effectiveness of groynes has been researched at length (ref. 3&7). Undoubtedly they perform a useful function under some circumstances; in others they are positively harmful. Whatever their usefulness, they form a prominent visual feature of many of our coastlines and may be seen on picture post-cards from Southend-on-Sea to Sutton-on-Sea and from the turn of the century to the present day.

25. At many locations, groynes have been allowed to deteriorate. Where they perform usefully, no doubt they will be maintained but where they are not, then it is unlikely that scarce resources will be utilised to remove them. These abandoned groynes may constitute a hazard to holiday makers (particularly the more adventurous children) and to navigation. There is a fine distinction to be drawn between something (such as an unfenced sea wall) which is inherently dangerous but is "there" and something (such as the jagged end of a corroded steel pile) which is potentially lethal and should be removed.

26. Bitumen groynes (such as those at Skegness) which have outlived their usefulness can disintegrate into unsightly and unsafe lumps spread over a holiday beach.

27. All of the features described above are man-made. It could be said that if we choose to tamper with the elements by building things on the beach, then we must expect to take the consequences when they go wrong or wear out. Those examples which follow are the results of natural processes over which we have little or no control. We can only hope to ameliorate their effects. The consequences of these natural changes for our coastal activities are potentially far greater than the effects of deterioration in our own creations.

SALTINGS

28. About a quarter of the region's sea defences are protected on the seaward side by salt marshes.

29. In some areas, notably The Wash, the general trend is accretion. Indeed, the process has been going on for many centuries and successive generations have reclaimed land in strips about 500m wide as the marsh grows. At some locations (e.g. Wainfleet) as many as four lines of defence are still visible with archaeological evidence of several more. Is this "deterioration"? Since the whole of The Wash is designated a National Nature Reserve and since conservation interests have prompted a moratorium on reclamation schemes, it is to be presumed that, from this point of view at least, seaward encroachment of the shoreline is "a bad thing". Be that as it may, there is no call for coastal engineering works in this part of The Wash and this paper will take no further cognisance of it.

30. By contrast, the Essex saltmarshes are eroding in a manner which gives rise to serious concern. The rate of

retreat at the edge of the saltings is as much as 3 m/yr. and the erosion of saltings is taking place over a significant frontage. This state of affairs presents a serious threat to the coastline proper since every metre of saltings lost permits the generation of slightly greater waves at the defence line. Many of these defences (which, remember, are made of clay and may be badly fissured) have no revetment at present. It follows that any increase in wave action is dangerous and could create progressive erosion leading to a breach (ref.4&5).

31. The problem is compounded by the relatively low value of the land protected. The cost of reconstructing these defences, or even refurbishing with revetment to counteract the loss of saltings could well be greater than the capitalised benefit of doing so. In these circumstances it is necessary to think the unthinkable. Should part of "this sceptred isle" be given up to the sea?

32. Clearly, any feasible stratagem to arrest or, better still, to reverse the erosive process is worth considering. Following successful experimental work carried out in Schleswig-Holstein, polders of various shapes and sizes have been constructed at Dengie and Mersea Island. Various materials (e.g. brushwood, hardwood off-cuts, nylon mesh), have been tried for the separators (ref.5). The results in some areas have been encouraging with positive regeneration of salting growth. The experiment continues.

33. Similar problems of salting erosion occur in The Humber, albeit on a smaller scale.

FORESHORE LOWERING

34. All of the processes described up to this point are overshadowed by the lowering of significant lengths of foreshore. The evidence of the Halcrow study (ref.6) indicates that the situation is getting worse and predicted trends (notably the rise in sea level) are likely to exacerbate the difficulty.

35. Following the CIRIA Groynes Study, (ref.3) Lincoln Division of Anglia embarked on a further study of groyne behaviour along the exposed North Sea frontage of the Lincolnshire coast between Mablethorpe and Skegness. The Division needed to make policy decisions for the management of its 262 timber groynes along this frontage. The CIRIA study was not site-specific and in order to apply locally the general conclusions derived nationally, existing data were analysed, new data collected and the behaviour of the coastline studied. The results have a bearing on the subject of this paper since a large part of the work concerned itself with the movement of beach material (ref.7).

36. Following the 1953 event, a series of beach cross-sections, at 18 locations, was taken and this survey

has been repeated (with some gaps) at monthly intervals up to the present time. The resulting data set has been studied by Hydraulics Research (ref.8), formed a key part of the Lincolnshire study and has now been absorbed into the Halcrow study. It is said to be one of the most extensive sets of data on beach behaviour in the world.

37. Eleven of the cross sections were selected for close analysis and, of these, three are selected here to illustrate the very different behaviour of beaches only a few kilometres apart. Table 1 shows net changes in beach level from 1959 to 1985. Expressed in a different form, Table 2 shows variations over significant periods based on three-year mean values.

Table 1. Net beach level changes September 1959 to April 1985 (m/yr)

Site	Distance offshore (m)						
	0	20	40	60	80	100	120
Mablethorpe	-0.03	-0.03	-0.03	-0.02	-0.02	-0.02	-0.01
Trusthorpe	-0.02	-0.02	-0.02	-0.02	-0.01	-0.01	-0.01
Anderby Creek	+0.03	+0.03	+0.03	+0.02	0	0	+0.01

Table 2. Average beach level variations (m/yr) over 0-100m width of beach, based on 3-year means

Site			
Mablethorpe	0 (1962-74)	-0.10 (1974-80)	+0.04 (1980-83)
Trusthorpe	-0.01 (1962-74)	-0.07 (1974-80)	-0.07 (1980-83)
Anderby Creek	0 (1965-78)	+0.10 (1977-83)	

38. Mablethorpe C.E. (Convalescent Home-since demolished) is one of the frontages where a consistent lowering has taken place from a relatively high base (average level; 3.6m O.D. over 25 years). Trusthorpe is one of the lowest beaches on the Lincolnshire coast (average level 0.7m O.D. over 25 years) and getting lower. Anderby Creek is high (average level 4.0m O.D. over 25 years) and getting higher.

39. Even in moderate conditions, there is a striking difference between the wave action observed at Trusthorpe and that a few kilometres away where the beach levels are higher. Colleagues visiting this location during storm

COASTAL STUDIES

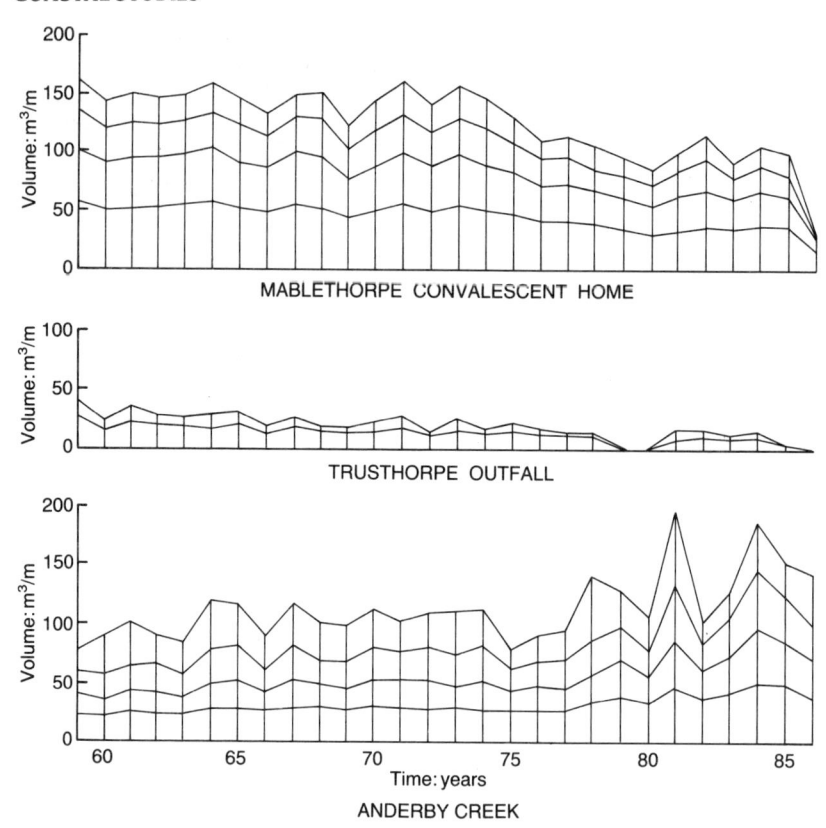

Fig. 4. Volume of sand above clay layer - annual averages

conditions speak of a very real sense of fear generated by the combined effects of noise, spray and the movement of the ground under wave impact.

Figure 4 illustrates the volume of sand on each of these three beaches from 1959 to 1986. Figure 5 illustrates the same data analysed in the form of three year means. The trends speak for themselves.

RETREAT

40. The Lincolnshire study was undertaken in connection with groynes, but it quickly found itself growing into a wider study of coastal behaviour. It became apparent towards the end (late 1986) that further work was needed to arrive at informed decisions on the management of the coastline. Around the same time, awareness was growing of similar needs to manage the sea defences of the whole of the Anglian coastline. So began the Halcrow study.

41. The data relating to the general lowering of beaches in Lincolnshire was refined by Halcrows and used in conjunction with similar (albeit less extensive) data

PAPER 12: WAKELIN

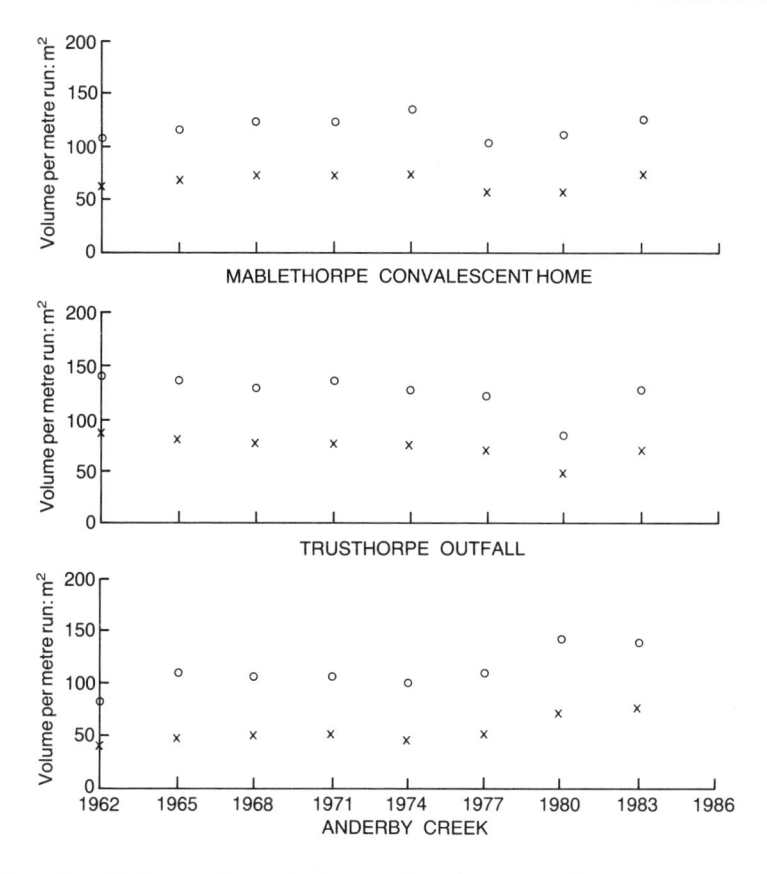

Fig. 5. Volume of sand above clay layer - three-year averages

gleaned from other frontages in Anglia. Following from this, an examination was carried out of the retreat of the Anglian coastline as revealed by a study of high and low water marks on Ordnance Survey maps.

42. The earliest versions of the 6 Inch to One Mile series of maps date from between 1850 and 1890. The latest revisions of the maps (at 1:10,000) date from the 1970's.

43. The values therefore represent the mean rate of retreat over approximately the last 100 years. Separate values were calculated for the high water mark, the low water mark and, lastly, the coastline as a whole: all this at 250m intervals. These values were then averaged over 1km lengths of coast and the mean value assigned to the mid-point of each length. Where the low water mark is too far offshore and not covered by the O.S. maps, only the high water mark and the coastline are shown.

44. A summary of the results of this analysis (in terms of the mean retreat in m/yr over broad stretches of the coast) is shown in Table 3. It is significant that the overall rate of retreat of the whole coastline is 280mm/yr.

145

Table 3. Summary of Coastal Retreat Rates (m/yr)

		Coast-line	HWM	LWH
Flamborough to Bridlington length = 10 km	mean s.dev	0.10 0.09	-0.0 0.17	0.25 0.22
Bridlington to Easington length = 60 km	mean s.dev	1.14 0.54	1.04 0.57	1.95 0.64
Easington to Spurn length = 4 km	mean s.dev	0.00 0.00	-0.42 0.34	0.65 0.29
Humber to Gibralter Point length = 48 km	mean s.dev	0.14 0.98	-2.19 3.28	0.25 1.58
The Wash to Weybourne Hope length = 56 km	mean s.dev	0.23 2.07	-0.21 1.49	1.41 2.36
Weybourne Hope to Cart Gap length = 35 km	mean s.dev	0.59 0.47	0.50 0.50	1.00 0.59
Cart Gap to Yare Mouth length = 28 km	mean s.dev	-0.08 0.77	-0.26 0.99	0.10 1.03
Gorleston to Slaughden length = 50 km	mean s.dev	0.68 1.11	0.75 1.17	0.94 2.27
Slaughden to Orfordness Spit length = 16 km	mean s.dev	n/a n/a	0.20 0.52	0.27 0.50
Shingle Sheet to Haven Ports length = 16 km	mean s.dev	0.14 0.50	0.00 0.56	0.40 0.78
Haven Ports to Hamford Water length = 5 km	mean s.dev	0.08 0.40	0.87 1.09	3.67 1.50
Hamford Water to River Blackwater length = 33 km	mean s.dev	0.03 0.36	0.39 0.80	2.02 1.57
Overall Mean		0.28	0.061	1.076

Notes: retreat is positive
s.dev = standard deviation

45. That the work has been done for the whole of the coastline will be seen from the table. For comparison with the earlier figures, the detailed results (showing the rate of advance/retreat for each kilometre) are selected showing the same Lincolnshire frontage. These are shown on Figure 6.

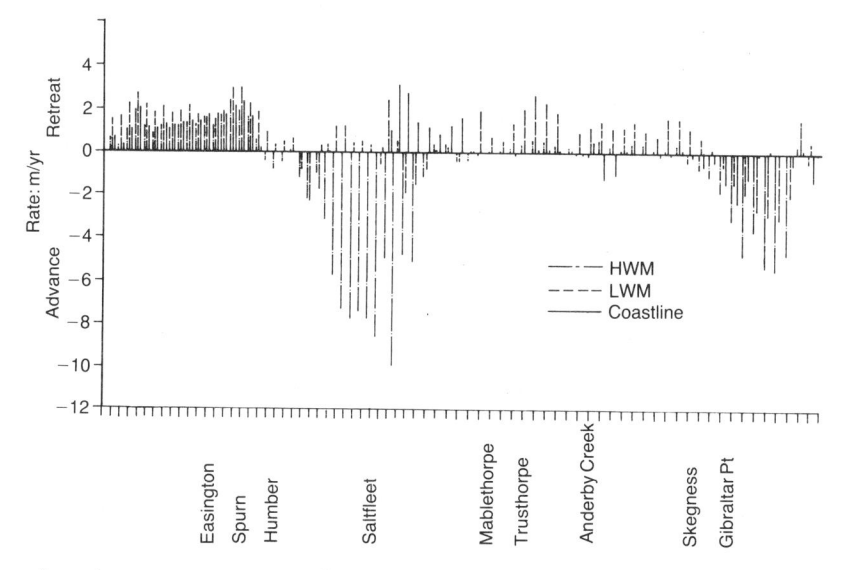

Fig. 6. Beach advance/retreat along Lincolnshire coast measured over approximately 100 years

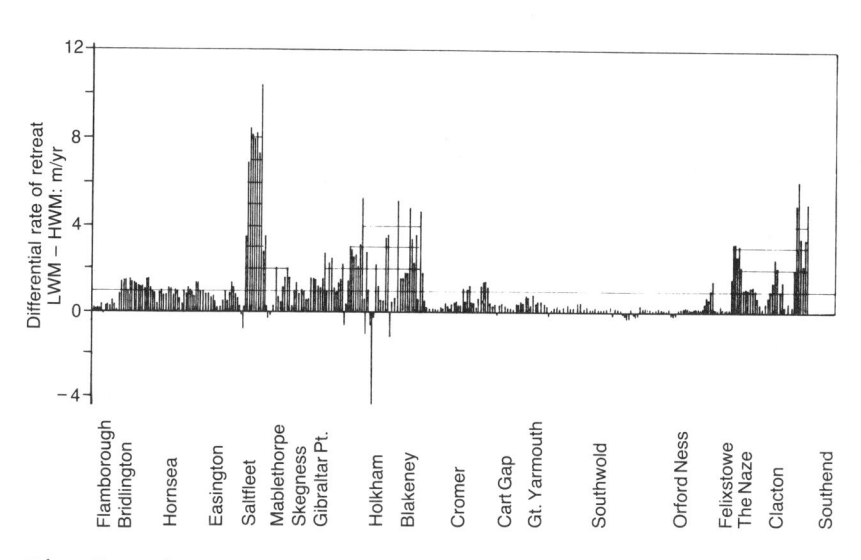

Fig. 7. Differential beach retreat

COASTAL STUDIES

46. The high and consistent retreat of the North Humberside coastline will be noted. The whole of this coastline is at a relatively high level and it is immediately apparent to the visitor that it is indeed eroding. Houses near the cliff top are clearly uninhabitable. That the analysis shows 2m/yr is not surprising.

47. Once again, we have to ask is this "deterioration"? Put the question to the owner of one of the doomed houses and no doubt he will express his affirmative opinion with great forcefulness. And yet; the erosion of these cliffs provides an essential supply of beach material for beaches farther South. We begin to venture into socio-political areas of consideration but from the point of view of good coastal management, there can surely be no doubt that these cliffs should be allowed to continue to erode.

48. Other results of this analysis also bear out first hand experience of the behaviour of this coastline. The dune frontages around Saltfleet and at Gibralter Point are accreting. The Mablethorpe to Skegness frontage is "above the line" i.e. retreating. It will be seen, however, that the "differential retreat" (i.e. the difference between the rate of movement of the low water mark and that of the high water mark) is much more pronounced than the equivalent figure for the North Humberside coastline.

STEEPENING

49. The above analysis introduces the concept of foreshore steepening. It is clear from a study of these data, assembled for the first time in the Halcrow study, that the average gradient of the vast majority of the Anglian coastline is becoming progressively steeper with time. Halcrows draw intriguing conclusions from correlations between this and other properties of the coastline. Sufficient to note here that the phenomenon exists and occurs over a very substantial proportion of the coastline.

50. If the Humber, The Wash and the Essex estuaries (i.e. the more sheltered frontages) are excluded, 71% of the coastline is experiencing retreat. Within this length 55% (i.e. two thirds of it) is also undergoing the process of steepening. In addition to that, where there is no lateral movement, a further 11% of the coastline is steepening. This is illustrated for the region in Figure 7.

51. If these trends continue, it is not going to be good enough simply to build bigger and better sea walls; they would eventually be undermined. An alternative, more fundamental, way must be found to mount a counter-offensive.

DOWN-DRIFT MIGRATION

52. In places where littoral drift in one direction is dominant, large quantities of beach material will move and the coast will change its shape. Naturally. If we have built communities or for some other reason are unwilling to relinquish land which would otherwise be lost by this process, then the remedial action necessary to preserve the "status quo" will usually be large in scale and therefore expensive.

53. If the value of the property we wish to preserve is relatively low (or if we have difficulty in evaluating it) then we must answer some difficult questions before concluding our assessment of worthwhileness. Aldeburgh is a case in point. Orford Ness is moving. Naturally. If nature is allowed to take its course, the spit will breach, agricultural land will revert to tidal marsh and the much valued (although not evaluated) sailing facilties in the R.Alde at Aldeburgh may be lost to the detriment of the town (ref.9).

54. Another example is on the Hunstanton to Snettisham frontage in The Wash where the Southerly drift has exposed the toe of the stepwork at Heacham. (Fig. 8). The engineering solution appears to be to nourish the up-drift (i.e. Northerly) end of the frontage with shingle transported from the down-drift end. This solution is not regarded with favour by the District Council who are mindful of the town's income from holidaymakers with a preference for sand. Once again, we are in the realms of beach management rather than (or in this case as well as) building bigger and better defences.

DEEP WATER CHANNEL MIGRATION

55. In the North of the region, on the Humber bank, sea defence toes are being undermined by deep water channels migrating from one side of the estuary to the other.

56. At one location, in the vicinity of Stallingborough, research has shown that the main deep water channel has been moving steadily Southwards since the turn of the century and that it will continue to do so 'until the Estuary reaches maturity" (ref.10). In the meantime, the seaward toe of the sea defence has had to be extended progressively downwards. No less than six capital schemes of this kind have been undertaken since 1958 and now, over a distance of some 2km further works are urgently required.

57. The proposed solution is to construct open stone ashalt revetment down to a depth which will prevent undermining of the toe as the channel continues its inexorable migration.

DUNE MIGRATION

58. Everyone agrees that sand dunes are one of the best forms of sea defence. If only we could encourage them to develop wherever tidal flooding threatens, our problems would be over. They cost almost nothing to maintain and provide excellent habitats for wildlife and holiday-makers (although not usually at the same place).

59. However, dune sand does move, like beach material, under the influence of natural forces and where this gives us human beings problems, we call it "deterioration". Dune defences North of Great Yarmouth cannot be maintained to provide a full defence for Broadland and hard defences have been built. The interface between the dune and the sea wall presents a discontinuity which Nature abhors. The wall cannot move but the dunes do and, on occasion, the wall has had to be extended to tie back into the shifting dune.

FUTURE MANAGEMENT

60. What will tomorrow's coastal engineering works look like? Since we have created assets along our coast and are reluctant to give them up, we are committed to defending the "status quo" over a large proportion of the total frontage. This represents an unnatural state of affairs and the struggle will get progressively harder unless we can devise ways to make nature work with us rather than against us.

61. We must become more subtle in our engineering with less emphasis on steel sheet piles and reinforced concrete; more on the encouragement of movement of material, be it in dunes, offshore or on beaches. In other words, we must learn to <u>redirect</u> the "great sources of power in nature" rather than opposing them. Like judo.

62. Where we are committed to hard defences, we must learn to extend their lives. Those defences built after 1953 have largely reached the ends of their useful lives. Bearing in mind the urgency with which they were built and the hostility of their environment, they have lasted well. However, despite improvements in construction methods, it seems likely that the defences we are building now may not last much longer unless we can find some way to defend <u>them.</u> The best way is to bury them. Preliminary work shows that the cost of even large scale beach management work may be justified solely by the extension of asset life.

63. The appraisal of future capital projects must take a broader view than in the past. This is not to criticize present methods of appraisal; only to say that we must make use of the wider knowledge which is becoming available. An action on one part of a coastline (or off it) may well have a profound effect on another part many kilometres distant.

Fig. 8. Exposure of stepwork toe Hunstanton South Beach, 1987

64. Apart from a general "softening" of our approach, we shall undoubtedly see other changes in emphasis. Artificial headlands (such as at Clacton) will have a high visual impact, off-shore breakwaters less so. All of these techniques will have as their objective the encouragement of beach building.

65. Another change must be in our attitude to the location of the coastline. What is there now is very different from what was there a century ago; and different again from what will be inherited by the next generation. The reader will not have found the word "abandon" in this paper so far. Before long we may have to come to terms with the concept of losing significant areas of land which cannot economically be defended.

66. Only by (a) increasing our understanding of the natural process at work and (b) adopting policies of regional coastal management shall we be able to achieve a proper balance between the various economic, social and environmental pressures over the decades to come.

ACKNOWLEDGEMENT

67. The Author wishes to thank the Directors of Anglian Water and gratefully acknowledges the assistance by his colleagues. The opinions expressed are those of the Author and do not necessarily reflect Authority policy.

REFERENCES

1. ANGLIAN WATER. Survey of land drainage needs, 1978. [Water Act 1973 S.24(5)].
2. CIRIA. TN124 Maintenance of coastal revetments.
3. CIRIA. Research Project 310: Effectiveness of groyne systems.
4. HYDRAULICS RESEARCH LTD. Report SR170 1988. Essex saltings.

5. MASCALL A. Essex Saltings; their loss and restoration 1987.
6. HALCROW & PARTNERS. Sea defence management study for Anglian Water, 1988.
7. ANGLIAN WATER. Historical review of the performance of groynes on the Lincolnshire Coast, 1987.
8. BRAMPTON & BEVAN. Beach changes along the coast of Lincolnshire 1959-1985 H.R.L. 1987.
9. ANGLIAN WATER. Aldeburgh Sea Defences. Engineer's report, 1988.
10. HALCROW & PARTNERS. Humber bank study; Stallingborough N. Beck to Oldfleet Drain, 1988.

13. The Anglian Sea Defence Management Study

C. A. FLEMING, PhD, MICE, PEng, Sir William Halcrow & Partners Ltd

---

SYNOPSIS. A major coastal management study for the entire Anglian Region has been initiated. Studies to date have included the collation of a vast amount of different types of data relating to all aspects of coastline management. These have been analysed through a relational database mapping system and have revealed a number of significant findings. A short term management strategy has been developed pending the outcome of further studies.

INTRODUCTION

1. There was a major reconstruction of the sea defences along the east coast of England following the 1953 floods. These defences have been maintained, extended and rebuilt during the ensuing years. Anglian Water who are responsible for one of the longest and most vulnerable coastlines in Britain, stretching from the estuaries of the Humber to the Thames decided that it was time to re-think the sea defence system as a whole and make some major reinvestments for the security of the coastline in the future. They therefore commissioned a study of the foreshore which could help them establish a coastal management strategy based on the latest data and analytical techniques available. Thus the Anglian Coastal Management Project has emerged and is probably the most extensive study of coastline properties and processes to have been carried out in the UK in recent times.

2. The general terms of reference for the investigations have been to provide an understanding of the mechanisms causing changes in foreshore levels along the Anglian Coast and to make recommendations for the future design and management of sea defences. Other issues related to the structural integrity of existing coastal works or economic and soci-political matters were outside the scope of the study.

3. One of the primary activities in the first phase of the project has been the collection of data from historical sources and studies as well as from present day observations, and the setting up of a new type of highly specialised computer relational database linked to a geographic mapping

system in order to obtain a working management structure. This has been a major exercise requiring the compilation; screening and verification of numerous different data sets providing key information of a wide variety of scales.

4. Data on nineteen primary variables have been collected for the 100 km or so of East Anglian coastline. These include wind, wave energy, water levels, currents, geology, morphology, coastal works, infrastructure, ecology, rainfall etc. Information from aerial photographs, historic records and records of works carried out have also been used to provide data. Additional studies have included long term hindcasting of the deep water wave climate, wave refraction modelling, a study of secular trends in sea levels as well as geological history and tectonic movement.

METHOD OF APPROACH

5. The coastline is an extremely complex interface between the land and the sea. Changes may take place at many different timescales and over many different spatial scales. A strategic approach to coastline control and management requires such changes for a coastal regime or region to be understood. Methods of solution to erosion problems may be indirect as opposed to the more traditional approach, where seawalls or revetments are sited as a means of satisfying hinterland requirements. Thus, in order to develop a coastline management approach it is necessary for the natural processes and man-made controls to be examined at both regional and local scales.

6. A large number of primary and sub-variables need to be considered (see Table 1). For any reasonable length of coast line a very large volume of data may be collected or generated and it must be adequately catalogued and be easy to extract for analysis. Thus, the use of suitable databases for the control and manipulation of data becomes necessary. In reviewing previous work related to coastal processes around Anglia it was clear that here was already a large volume of information available. The rational of the proposed approach was to extract a coherent picture from the available data by means of:

- development both a referral and a relational database,
- instigation of a number of supplementary studies,
- carrying out a regional interpretation of the available data to determine individual coastal units,
- development of short term management options,
- consideration of further strategic investigations required to establish a long term management strategy,

7. Anglian Water's sea defences run from Canvey Island on the Thames to Trent Falls on the Humber Estuary. Given that the objective of the study was to establish management procedures based on a sound regional understanding, it was felt necessary to extend the northern boundary of the study

area to incorporate the Holderness coast. This thereby includes an important sediment source for the East coast. In contrast the Kent coast, on the south side of the Thames Estuary, was not considered to interact significantly with the regime along the Essex coast and was not therefore included. The extent of the study area is shown in Figure 1.

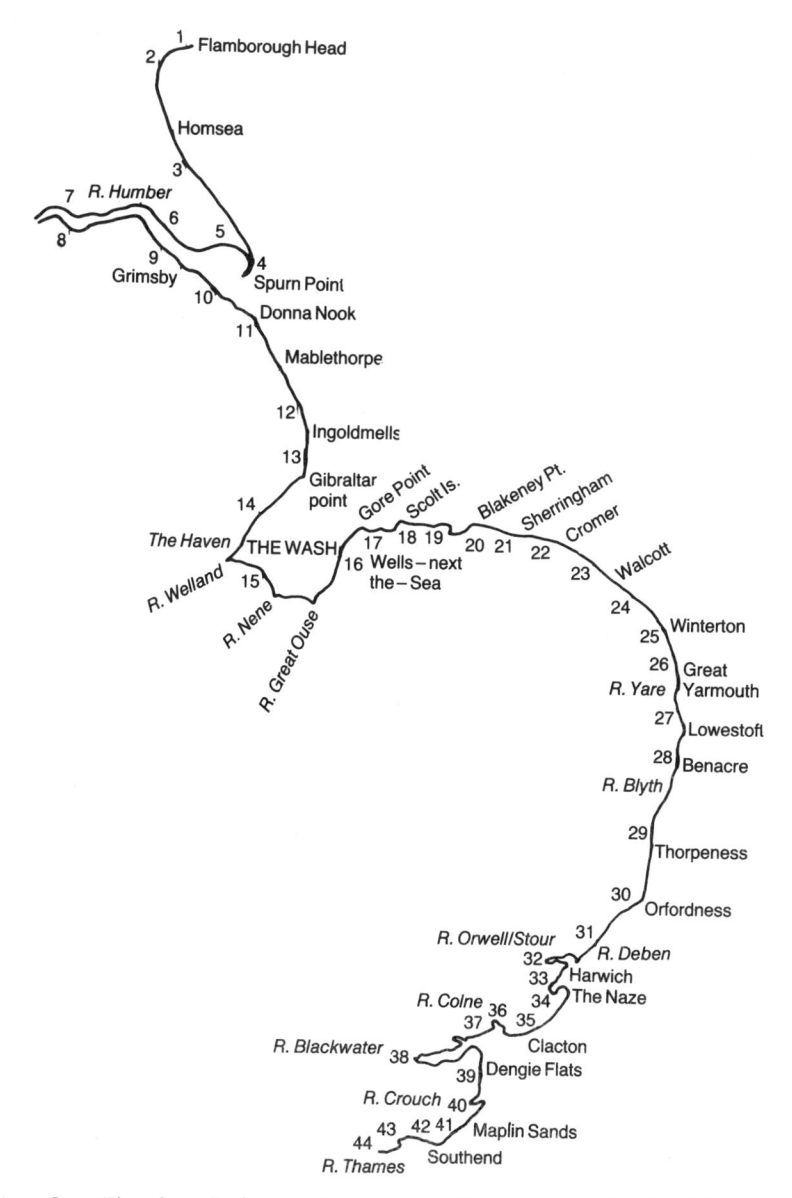

Fig. 1. The Sea Defence Management Study for the Anglian Region - coastal units

COASTAL STUDIES

REFERRAL DATABASE
8. This database essentially provides a listing of relevant references and data sources. It includes entries for reports, papers and relevant literature, as well as details of the type, extent and holder for the various sources of measured data that have been identified.

9. The database was set up using a proprietory software package and has been constructed using two separate record structures, one for references and another for environmental data. The database can be searched on a number of fields, such as key word, data type, author, coastal region, location and so on. Searches can be constructed in several ways to give either a comprehensive listing, such as all references for a particular region, or much more specific responses such as data of a particular type, in a particular region and of a particular time frame. It therefore provides a rapid means of identifying what data and information is available.

SUPPLEMENTARY STUDIES
10. It was apparent that a number of important variables could not be adequately defined by using existing sources. Six supplementary studies were undertaken with the objective of providing additional data or some specific aspects of further understanding. These studies are briefly described in the following sub-sections.

Wave Climate
11. A definition of the wave climate offshore and at approximately 10km intervals along the coast was required. The offshore area was initially sub-divided into six areas centred appoximately on Skegness, Cromer, Yarmouth, Dunwich, Felixstowe and Clacton. The deep water wave climate was then generated at hourly intervals in the form of hindcasts of wave heights, periods and directions. Each hindcast data set was plotted as time series and probability distributions to validate against any existing measured data.

12. A multiple grid spectral wave refraction model has been constructed for the whole of the east coast offshore area. This was used to determine the nearshore wave climate at 38 points along the coast, spaced at approximately 10km intervals. Time series of offshore waves and tidal water levels were used with the results from the refraction models to derive the inshore time series at each point. Good agreement with measured data was found when shallow water limits to the energy spectrum were applied. The nearshore data was also used to derive statistics for alongshore and onshore-offshore wave energy by direction, season and annually. In general there appeared to be a good qualitative agreement between known shoreline processes and the prevailing wave conditions predicted by the model.

Current Residuals

13. Some clarification of the coupling between the residual flow regime in the Southern North Sea and local wind effects in determining nearshore current residuals was required. The key feature of water movements in the area is the consistency of a north-easterly residual stream offshore South-East Anglia and an easterly residual stream off North Anglia. A separation zone can be both predicted and observed, offshore the Lowestoft, Great Yarmouth region. This can be distorted under strong wind conditions.

Beach Profile Analysis

14. It was necessary to establish long term retreat rates using both measured profile data and historic Ordnance Survey maps. A large amount of beach profile data has been identified and gathered for storage on one large database. As an example there are more than twenty years of data at quarterly intervals at fifteen sites on the Lincolnshire coastline.

15. Statistical analysis on the profiles was undertaken to identify trends in both the horizontal movement and the slope of the beach face at Mean High Water, Mean Sea Level and Mean Low Water Levels. The results were somewhat erratic and indicate the need for some guidelines on the extent and frequency of future survey work. Nevertheless where data is reliable some clear trends do occur in some cases. it is also apparent that, in the vast majority of cases, beach profiles do not extend far enough offshore to pick up profile movements on the lower profile dam to closure depth. This is not surprising due to the difficulties with surveying below low water.

16. Coastline retreat data was also derived from historic Ordnance Survey Maps. The rates of retreat were calculated from the movement of the tidal marks and coastline between the first edition of the 6 inch/1:10,000 Ordnance Survey maps and the latest revision. For most areas the first edition dates from between 1850 and 1890 and the latest revision dates from the 1970's. The values therefore represent the mean rate of retreat over approximately 100 years. Separate values were taken off for the coastline, high water mark and low water mark at 250m intervals. These were then averaged over 1 km. The data has all been entered into the Relational Database.

Extreme Water Levels

17. Extreme still water levels for twenty locations from the Thames Estuary to the Humber have been estimated using the Generalised Extreme Value analysis of recorded annual maximum levels. However, the data sets were of varying length. Data was analysed and adjusted for secular trends, but the results were inconsistent. A number of different extreme value analyses were also used and the recommended method was found to be in good agreement with other estimates.

Sea Level Change

18. A perspective on sea level change in a geological setting was required. Studies have established whilst there have been changes, at some time reversal of trends have occurred over periods of tens of decades. Contemporary tide gauge data show an increasing trend and that this is likely to be accentuated by global warming and ice melting due to the CO2-effect. It can also be shown that coastlines react in other ways than retreat due to flooding. It was not found possible to quantitatively differentiate between changes in regional eustatic sea-level and tectonic movements. However, regional sea-level comtemporary trends indicate that levels are rising by about 2mm/year along the coast of East Anglia increasing to 4mm/year south of the River Blackwater.

19. A review of various aspects of future changes in global sea levels was carried out so that some reasonable measure on the widely varying estimates of long term rise could be made. Estimates for the next thirty years or so are relatively modest, but there is a fairly high degree of uncertainty.

Literature Review

20. A review was carried out to ensure that all of recent literature relating to coastal processes relevant to the Anglian coastline had been taken into account. Over one hundred publications were identified as being relevant. The subject areas were divided according to coastal domain (i.e. fully developed sand beaches, cohesive shores, salt marshes and dunes and offshore features) and thence by time-scale (i.e. tidal, single event, seasonal and geological). A further sub-division was made through natural coastal processes and man's influence on coastal processes.

RELATIONAL DATABASE

21. A relational database capable of storing the spatial distribution of a large number of data sets was adopted. The database can be interrogated both spacially and analytically, enabling relationships between variables to be sought, and so providing insight as to the governing coastal mechanisms.

22. The relational database was used to meet the following objectives:

i) to map relevant variables for the entire coastal region
ii) to use the graphical output of the system to present each variable or combination of variables on a series of maps
iii) to assess the inter-relationships among variables and their contribution to coastal erosion.
iv) to produce interpretive maps which form the basis of a coastal management policy.

23. Some 60 different agencies were approached with regard to collecting data. It was concluded that there was sufficient data to provide at least a basic description of

each variable, although in a few cases this was limited to a relatively coarse regional description.

24. The software used was developed for use on a microcomputer with two screens, one for text and another for graphics display. The configuration for this project was established to provide facilities to create, display, edit, report, plot and query the data. Two forms of query facility are available. The first provides the ability to make enquiries of a single variable (e.g. waves, or morphology, or coastal works, etc). The second enables the results for a number of enquiries, using different variables to be combined (e.g. waves and morphology and coastal works, etc). This is achieved by plotting the result of each enquiry as an offset line from the "coastal reference string", thereby permitting the spatial coincidence of different fields within the database to be rapidly tested.

Definition of Variables

25. In all, 19 main variables have been included in the relational database. These were selected on the basis that they either provide information on the direct influences and responses of the coast (e.g. waves, coastal morphology, rate of retreat, etc) or on their implications with respect to the impact of the erosion and any defence strategy that may be implemented (e.g. present coastal works, SSSIs, land use, etc). A complete list of the main variables, together with a summary of their significance is given in Table 1.

TABLE 1: Significance of Main Variables

Main Variable	Significance
AGRICULTURE	- changes to habitat - drainage patterns and run-off
CURRENTS	- controls sdiment movement in offshore zone - links nearshore processes with far field effects in the North Sea
COASTAL MOVEMENT	- indicates areas of high/low activity - necessary to be able to make forecasts - relates to sediment budget
COASTAL WORKS	- required to establish any interaction with coastal processes
ECOLOGY	- a measure of shoreline (cliff, dune, saltmarsh) stability, shelter, relationship to rivers and estuaries - required to assess environmental impact

COASTAL STUDIES

FISHERIES	- changes in habitat
SEDIMENTS	- determines mobility of material - can help to establish source(s) of material - basis of sediment budget
WAVES	- determines potential for shoreline erosion and accretion - influences movement and height of offshore banks
INFRASTRUCTURE	- constrain the coastline - can alter the inshore wave energy regime
INDUSTRY	- coastal impact (eg outfalls) - threat to habitats - modify sediment paths - alter local currents, wave conditions - lead to changes in the sediment budget
JURISDICTION	- important for the development of management strategy
WATER LEVELS	- major effect on coastal processes - controls extent of wave influence on shoreline - relates to potential for land flooding
MORPHOLOGY	- provides basic description of coastline - features can have a physical significance (eg offshore banks dissipate wave energy, cliffs can provide a sediment supply, etc) - widths of the foreshore provide an indication of plan effects - slopes control the form of incoming waves - indicate nature of sediment transport - represent sediment sources and sinks - intertidal features indicate beach cycles and on-offshore movement
BIRDS	- required to assess environmental impact

CONSERVATION SITES	- special consideration to prevent undesirable changes
WATER QUALITY	- influences vegetation and hence shoreline stability - can effect marine life and alter habitats - density effects influence and transport regimes
RAINFALL	- influences ground water levels and river discharges - relates to the sediment load in rivers - can effect cliff stability
TEMPERATURE	- seasonal variations may contribute to erosion
WIND	- generates waves and surges - governs sub-aerial erosion and deposition

26. Once data became available for a particular variable it was possible to consider what fields should be included in the database. For a field using descriptive information a standard classification system has been developed to ensure that consistent definitions were always used. Thus for each field there is a fixed number of classes. For example, the classes for the field 'Relative Position' are: Hinterland, Backshore, Foreshore, Nearshore, Offshore. These are defined in Figure 2.

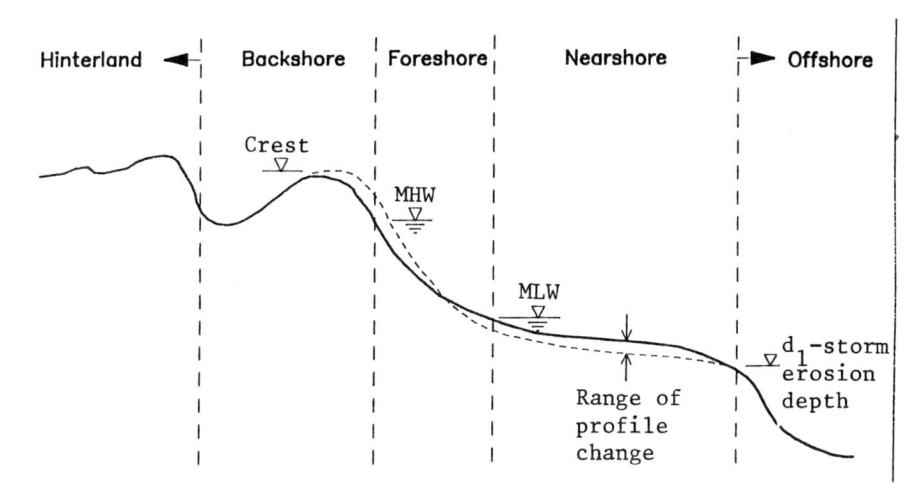

FIGURE 2: The Sea Defence Management Study for the Anglian Region (Sir William Halcrow & Partners Ltd)

27. Whilst the development of many of the classification systems was straightforward, other required more careful consideration. Some descriptions varied from one locality to another and a common nomenclature had to be evolved. This was often achieved by a process of iteration, involving re-mapping the variable several times.

28. A further important aspect was the need to consider how the variable could usefully be interrogated using the mapping capability of the system. For example, the coastal works classification system provides for this principally through the 'position' and 'purpose' fields, with the other fields denoting a reasonably literal description. Another example is provided by the classification system for the coastal infrastructure. Here the first three fields, 'position', 'orientation' and 'description', are used to describe attributes which can be mapped and are potentially relevant to coastal zone processes. The remaining two fields then provide a more specific breakdown of the items and the possibility of examining associations with particular structural types (e.g. docks and harbours, outfall structures, etc).

29. The 'relative position' is particularly significant in that it allows all the attributes of the variable on a particular length of coast to be mapped. This is very relevant for the coastal morphology variable, for which a description is included against each class of the relative position field (i.e. hinterland, backshore etc). In this case the structure of the fields follows on from each class of the relative position field. In turn sub-fields each have their own sets of classes to provide the necessary description.

30. In addition to descriptions for each item, numeric data fields can also be added. Typically having provided a classification of an item, it is followed by some details, which might be in the form of dates, key dimensions, mean or extreme values, etc. The dimensions given are representative values for the length of the item as mapped and do not take account of localised variations. For instance, an item of coastal works is taken to be a length of a particular structure type over which there are no major changes of section, or a groyne field rather that the individual groynes. Other variables, such as winds and waves, are fronted by fields for 'direction' and 'season', and for each combination of the classes in these two fields data is assigned. This takes the form of mean annual conditions an extreme values. Thus, the record format varies from one variable to another due to the very different types of data that have been incorporated into the database.

ANALYSIS

31. The Relational Database can be used in a number of ways ranging from the relatively simple recovery of data, to the more complex research and interpretive techniques. In these studies most of the analytical effort followed a combination of houristic and inductive techniques. The former provided

classifications upon which interpretive maps were developed and which form part of the Anglian Coastal Management Atlas. The latter technique was used in conjunction with the derived classifications, to examine particular, known, problem areas and so develop appropriate management strategies.

32. The long term retreat data has been used to develop a retreat/advance classification system which characterises the coastal movement. The results show that some 70% of the exposed coastline has been subject to retreat during the past 100 years. Other areas of data interpretation have been wide ranging and a number of key insights have been found. These may be summarised:

- There are a number of different modes of coastline retreat that can be related to the geological setting.

- There appears to be some periodicity in coastal orientation which can be disturbed by the presence of shoreline structures.

- No correlation appears to exist between incident wave energy and long term retreat. Both onshore-offshore sediment movement and tidal currents appear to be responsible for significant sediment transport along the shoreline and, in effect, control the volume of beach supply.
- Different types of structures influence both the rate and type of erosion which takes place.

- A model for the development and movement of the Great Yarmouth banks has been derived.

33. Following on from the examination of governing processes, a more detailed interpretation at a local scale has enabled coastal units to be defined. The units comprise lengths of coast which exhibit coherent characteristics but are not necessarily independent of adjacent units.

SHORT TERM MANAGEMENT STRATEGY

34. The short term management strategy has been developed by relating Policy Options and the available understanding of coastal processes to propose Management/Engineering Options. The Policy Options are to -

- maintain existing line
- set back defence line
- retreat
- advance

35. The Management/Engineering Options required to implement the policy options in the course of time have been grouped under four main headings:

- do nothing
- reinstate
- modify
- create

36. These provide the basis for a short term management strategy, which have been designed to ensure consistency through the region. This should enable the initiation of a more strategic approach to the provision of sea 'defence Works'.

37. Notwithstanding the above, and in view of the substantial nature of some of the insights derived it is intended that a number of further investigations are to be carried out. These will involve the filling in of data gaps identified during the course of this stage of the project as well as refining conceptual models of offshore bank movement and beach response.

ACKNOWLEDGEMENTS

In the course of the project, the Staff of Anglian Water, its Divisions and the Marine District Councils have provided much assistance and valuable data. The project team was drawn from Sir William Halcrow & Partners Limited, British Maritime Technology and Hydraulics Research. Much of the material in this paper was provided by the project manager, Mr Ian Townend of Halcrow.

14. Implications of climatic change

K. M. CLAYTON, CBE, MSc, PhD, Professor of Environmental Sciences, University of East Anglia

SYNOPSIS. Starting from predictions of the future rate and amount of sea-level rise through to the middle of the twenty-first century, I list the adjustments available. Coastal adjustment to sea-level change, and the consequent management decisions that will be required are described for cliffs, drift-aligned coasts, dunes and saltmarshes. I conclude with comments on the timescale of change, on the cost of coastal engineering works at a time of sea-level rise, and on the timescale for changes in legislation and in attitudes.

THE OUTLOOK FOR THE COASTAL ZONE

1. Whatever may have been the outlook for our extensive coastal engineering works had sea-level remained constant, there is widespread agreement that the possibility of green-house-gas-induced sea-level rise threatens their long-term effectiveness. It is axiomatic that with rising sea-level, sea defences will be overtopped more frequently, and sites with falling beaches in front of sea-walls and other "hard" defences will suffer more severe onslaught during storms. These predictions are enhanced by suggestions that among the climatic changes brought about by greenhouse-gas warming will be increased storminess in our latitudes.

2. These simple consequences remain true whatever the amount and rate of sea-level rise, but the size of the problem and the time we have to adjust (and the choice of modes of adjustment open to us) varies with the amount of sea-level rise predicted. It may be reassuring to learn that the latest predictions by the Climatic Research Unit (CRU) at the University of East Anglia are at the lower end of the range for the period through to 2030 to 2050. Through careful analysis of the various factors involved, and the delays inherent in the buffering effect of the rate at which ocean temperatures (and thus ocean volume) can change, the CRU predicts a rise in the range 15-25 cm by 2030 and that it will take until at least 2080 for the rise to reach a metre.

3. However, even these modest increases in sea-level will have measurable impacts on coastal structures and the frequency of overtopping, while the rate of rise towards the end of the period is at least double that occurring today. For

some situations, the rate of rise can be as significant as the total amount of sea-level rise (ref. 1). The other problem is that the rise of sea-level follows a period when beach slopes have been steepening appreciably (Table 1). Indeed, the disparity between the movement of low tide and high tide marks along the North Sea coast south of Flamborough Head implies that 50 m of coastal retreat will have to occur if the profile of 1880 is to be re-established. Presumably this adjustment will occur in areas left to adjust to sea-level rise, and will be threatened even along protected coasts.

Table 1. Annual rates of recession of the coastline, HWMOT and LWMOT for representative sectors of the North Sea coast over the last 100 years or so. Values are $m.yr^{-1}$, negative values indicate progradation.

	Km	coastline mean	SD	HWMOT mean	SD	LWMOT mean	SD
Cliffs:							
Flamborough Chalk	8	0.10	0.09	0.00	0.19	0.24	0.23
Holderness till (N)	10	1.20	0.21	1.13	0.19	2.45	0.28
.. .. (C)	12	1.48	0.23	1.42	0.32	2.28	0.32
.. .. (S)	14	1.67	0.41	1.56	0.42	2.17	0.51
Norfolk till & sands	29	0.93	0.47	0.80	0.57	1.42	0.50
Covehithe sand/gravel	6	2.81	0.54	3.05	0.57	4.13	5.24
Low coasts:							
Lincolnshire	34	0.39	0.94	-2.06	3.64	0.74	1.23
North Norfolk	29	0.65	2.50	-0.64	1.83	1.29	3.07
East Norfolk	20	0.11	0.67	0.03	0.74	0.49	0.70
Shingle St/Felixtowe	16	0.14	0.50	0.00	0.56	0.40	0.78

4. The core of the message I have for coastal managers is that *provided the decisions are taken early enough and provided they are pursued consistently*, that adjustment to sea-level rise is not unduly difficult or unduly expensive. But effective - and in this I include cost-effective - adjustment requires a radically different approach from that followed today. I happen to believe that the inadequacies of our current approach to both coastal protection and sea defence are beginning to become apparent along difficult coasts such as those facing the North Sea, but there is no doubt that these inadequacies will surface very clearly as the rate of sea-level rise increases. So the message here is that *coastal management under conditions of rapid sea-level rise must embrace the whole coastal zone, it cannot as now be*

restricted to the beach.
5. This change of approach will involve considerable reorientation of the approach of coastal engineers who will have to evolve into - or be replaced by - coastal managers. It will involve changes of legislation appreciably more substantial than the tinkering outlined in the Green Paper of 1985 (ref. 2). It will also require a re-evaluation of the role of the various natural elements of the coast (beaches, sand dunes and salt marshes) and much greater emphasis on their ability to adjust naturally to changing coastal conditions. Too much of the coastal engineering of the past few decades has sought to stabilise coastal features to allow, or to preserve, development right up to high tide mark. In future we need to give these features room to move as they adjust to changing conditions, for if they are allowed to adjust, they will remain effective natural barriers buffering the power of the sea.

AVAILABLE ADJUSTMENTS TO COASTAL HAZARD
6. The first step in considering the managerial implications of sea-level rise is to recall the range of adaptations available to us. It is true that some of the adjustments listed are not at present covered by legislation, but we have time to remedy that:
   1) increase the height and/or strength of defences;
   2) improve beach levels through nourishment;
   3) move existing defences landward to allow roll-back of the active coastal zone;
   4) abandon defences and allow nature to take its course;
   5) prohibit further development in a "littoral hazard" zone and plan for existing properties to be abandoned in time;
   6) compensate owners of property in threatened areas to avoid the need for future protection;
   7) scale down the level of land use (e.g. return drained arable land to coastal marsh);
   8) flood-proof individual properties - or change their use.

It will be noted that engineering solutions to sea-level rise are but two choices of the eight listed here. Recently Bruun (ref. 3) has classified the adjustments possible into just three classes, stand-by, give way, and nature's course.
7. It is important to recall the timescale of the adjustments we are considering - something like 40-60 years or more; these are long in engineering terms. This relatively long timescale allows scope for more fundamental adjustments than are normally considered when discussing options in hazardous situations. It is long enough for many of the houses built since the last war to pass to the stage where they will either require major upgrading - if they are in locations or at a spacing which makes this type of renewal economic - or may reasonably be abandoned as at the end of

their economic life. It is long enough for changed policies on land use to have an impact, so that farmers may find an "economic window" within which they may more easily adjust to the need to take reclaimed marshland out of arable production, or allow drained meadows to revert to saltmarsh. Long-term planning of this type was attempted from the Town and Country Planning Act of 1947, and has gradually gone out of fashion as even the green belt is claimed as an anachronism. It may remain unfashionable over much of the UK, but there is a strong case for its readoption in the hazardous littoral zone.

LITTORAL ENVIRONMENTS AND SEA-LEVEL RISE

a) Cliffed coasts

8. There are three main types of cliffed coast in the British Isles, those of relatively weak rocks which are retreating at rates of 0.5 m/yr or more, the tougher rock cliffs standing above coastal platforms where rates are from 0.01 to 0.1 m/yr, and plunging cliffs with a low rate of retreat which may reach the rates of rock cliffs above coastal platforms. In passing we may note that while rates of retreat on the weaker rocks are becoming well known, little detailed work has been completed on the tougher rock cliffs. Indeed, the current view that rates of retreat are everywhere in the range 0-0.1 m/yr may turn out to be a misapprehension.

9. **Plunging cliffs** rise from relatively deep water and even large waves may strike them without breaking. Few measurements are available of their rate of retreat, though the assumption in the literature seems to be that they retreat even more slowly than cliffs in similar rocks which rise from a coastal platform. What does seem certain is that plunging cliffs will be little affected by a rise in sea-level - in terms of the depth of water adjacent to the cliff the change will be inconsequential. Hence if they pose so little problem at present that no-one has troubled to measure their rate of retreat, that happy state of affairs will continue.

10. **Tough rock cliffs rising from coastal platforms** are more problematic. If they are in equilibrium such that the rate of cliff retreat is linked to the rate of platform lowering (ref. 4), it is possible that the rate of retreat will increase as sea-level rises. That is to say the control over wave energy reaching the cliff will no longer solely be the rate at which the platform can be lowered, but that rate assisted by sea-level rise. Uncertainty about the response is whether the increasing water depth will allow continued and constant reduction of the platform, and whether the rate will change with increased water depth (if it slows down, cliff retreat could stay constant despite rising sea-level). If the constraint given sea-level rise becomes the resistance of the cliffs to wave attack, deeper water over the coastal

platform could again have no effect on the rate of cliff retreat. To these complications, we may add the fact that some geomorphologists suggest that cliffs and/or platforms may be relict features from past periglacial or interglacial periods, reoccupied by contemporary sea-level but not yet fully in equilibrium with today's conditions. If this is so, sea-level rise need not produce any change in response. Given all these uncertainties, it is fortunate that on the whole we are dealing with relatively tough rocks and thus relatively slow rates of retreat.

11. **Cliffs in weak rocks** (found mainly along the North Sea and English Channel coast) typically retreat at about 0.5 m/yr (Table 2), though rates of 2-3 m/yr are found for limited stretches and local rates of >5 m/yr occur, though these are probably not maintained for >50-100 yr. The

Table 2. Some retreat rates on cliffs in pre-Quaternary sediments. (in part after May, ref. 5)

	mean	max rates
Tertiary sediments	0.60 m. yr^{-1}	0.81 m.yr^{-1}
All chalk cliffs	0.21	0.51
Middle & Lower Cretaceous	0.43	0.49
Purbeck & Portland Beds	0.12	-
Jurassic mudrocks	0.42	0.43

spatial and temporal pattern of retreat of the Norfolk cliffs has been established by Cambers (ref. 6). She also showed that the rate of retreat is not limited by the need to maintain the cliff base free of sediment as the highest cliffs retreated at the highest rate. This situation is also true of chalk cliffs, thus under current conditions some other constraint limits the wave energy reaching the base of the cliffs and stabilises the rate of retreat. It is suggested that this is the rate of lowering of the offshore ramp plus the rate of rise of sea-level, if any.

12. If this deduction is correct (and it fits the long-term evolution of these cliffs well), then rising sea-level will speed up the evolution of these cliffs. The slope of the offshore ramp, and considerations of the overall geometry of the system, suggests that the rate of retreat will increase by about 0.35 m/yr for each 1 mm/yr rise of sea-level. Incidentally, this implies that contemporary rates of sea-level rise are not yet fully reflected in increased rates of cliff retreat, a suggestion supported by the observation that inter-tidal slopes have increased over the last 100 yr (Table 1). The upper limit to this retreat will be set by the rate at which sediment can be removed from the cliff base. This limit is probably at least 3m/yr for cliffs >40m high, and thus >6m/yr for 20m cliffs and $\geq$10m/yr for 10m cliffs. These high figures may be approached by the middle of the next century, especially as they are controlled by the rate of sea-level rise rather than the absolute amount.

13. **Existing rates of erosion pose considerable management problems on these weak-rock cliffed coasts**, as such areas as NE Norfolk, Holderness and Barton-on-Sea (Hampshire) show. The higher rates predicted (including the increase which must soon result from offshore steepening independent of future sea-level rise) can only be accommodated if immediate measures are taken to control development in the cliff-top zone under threat. If we envisage that new or rehabilitated housing should have a design life of 100 yr, this implies a zone 300m wide on high cliffs and up to 1km on low cliffs. Cost-benefit studies should be undertaken to decide in what circumstances the alternative policy of coastal protection might be feasible, but given the impact on beaches down-drift (see below), this would probably need to be achieved by beach feeding and this will not be cheap given the large volumes required to offset littoral drift. Of course sea-level rise will everywhere make beach nourishment a more attractive technique, and this will presumably lead to investment in more efficient techniques and thus lower costs.

b) Drift-aligned coasts

14. Considerable stretches of the British coast are drift-aligned in the sense of Davies (ref. 7) and fed by cliff erosion (the "feeder bluffs" of American authors) and in upland areas such as much of Scotland by sediment delivered to the coast by rivers. Many of these beaches are decorated by groynes, which in these situations can improve beach alignment and thus beach volumes, though this is not always achieved.

15. Detailed work remains to be completed on the adjustment of these beaches to sea-level rise, but it seems very possible that they will prove to be one of the naturally resilient parts of the coastal system, particularly where littoral drift volumes will rise as the feeder cliffs erode more rapidly. At the updrift end against the cliffs these beaches will retreat with the cliffs, but longer stretches will remain dominated by the natural alignment under littoral transport, while the larger sediment volumes will go some way (and perhaps all the way) in assisting an adjustment to sea-level rise. Models used on the drift-aligned barrier beaches of the Cape Hatteras coast show the important role of existing littoral drift volumes in modifying the two-dimensional Bruun rule adjustment to sea-level change (ref. 8) and in due course similar models should assist us in predicting the adjustment of drift-aligned beaches. I suspect that we shall find that higher and wider beaches will be formed from the increased sediment volumes, and that in combination with dune maintenance or even growth (see next section), these will be some of the most resilient coasts as sea-level rises.

16. The role of sediment supply in beach dynamics is supported by the relationship between beach steepening and coastal geology found along the North Sea coast. Wherever volume is lost through mud transport offshore, beach steepen-

ing has occurred. This is related to what Dean & Maurmeyer (ref. 9) call the "sediment compatability factor" (i.e. that retreat rates are directly proportional to the percentage of mud in shoreface-bypassed material). This relatively rosy forecast for the downdrift beaches is completely dependent on a management policy which allows the accelerated erosion of the cliffs from which the natural sand feed comes. Today most of these systems are stressed by attempts to defend the updrift cliffs with consequent reductions in sediment supply (ref. 10). It will be politically difficult to abandon attempts at coastal protection in these areas, especially as the rate of retreat will be increasing over current levels and cost/benefit criteria will become easier to meet. Yet apart from the difficulty of finding the money to defend all the British coast, wider cost/benefit analysis which took into account the downdrift beaches would identify the need for increased investment downdrift to compensate for reductions in sediment supply. The Green Paper of 1985 (ref. 2) makes some contribution towards this wider view by consolidating the separate responsibilities for high and low coasts. Nevertheless, the operation of a selective policy on coastal defences with the abandonment of long lengths of the coast to the natural balance of nature will require an major educational effort and strong national policies.

c) Dunes

17. The British show little respect for dunes, yet they are an important part of the long-term strength of a coastal system. While dunes are only well developed on relatively stable or even prograding coasts, as a natural store of sand capable of replenishment by natural processes, they can stabilise the beach system during storms and especially during storm surges. Once the additional sand then provided to the nearshore system has served its purpose, re-establishment of the normal beach slope will lead to the re-supply of sand to the dune system, rebuilding them over a period of a decade or more. All too often dunes are regarded by coastal engineers as a weak and ineffective barrier, a short-term view which undervalues their role as a sand store in time of need and their capacity (if not abused by trampling, and provided beach levels are maintained) to rebuild naturally without cost to the sea defence budget.

18. The Dutch have fully realised the value of sand dunes as a store of sand, capable with a wide beach of being built up (and rebuilt) naturally, yet providing an invaluable buffer of sand as erosion occurs during a prolonged storm surge. Indeed, they now regard a wide beach in front of a wide (e.g. 200 m) belt of dunes as the best protection for the flood bank behind. To this end they are prepared to feed low beaches to the point where they can supply surplus sand to increase the width of the dune system and the volume of sand in store. Few if any British coastal engineers seem to take the same view, and the normal British scene is to find

COASTAL STUDIES

the dunes over-run with the damaging tracks of holiday makers and scarred by blow-outs. Along the top of the beach there is often a wall, which inhibits the natural growth of foredunes and bars the sea from eroding the dunes during surges, so that their role as sand stores is not exploited. If sea-level is to rise, we shall need to adjust our perception of dunes as soft and irrelevant features along a coast, and instead seek ways to encourage their growth as a renewable buffer against storms and their associated surges. This raises questions of ownership and management in many dune areas.

d) Salt marshes

19. Man has been reclaiming coastal marshes by endyking for centuries, using the natural ability of the coastal system to build up the protected marshes by warping before finally excluding the sea on what he hopes will be a permanent basis. Indeed the gains of land area brought about by the transference of mud from eroding cliffs and river inputs to these prograding areas meant that at least until the early years of this century we gained more land from the sea each year than was lost.

20. While this process has continued in a few favourable areas such as the Wash, under conditions of rising sea-level it cannot continue. Indeed, it will be necessary to increase the already high expenditures on coastal banks and walls (which have already slowed the rate of reclamation) to keep salt water out of these reclaimed lands. The likely costs and benefits of a policy of raising flood banks to keep pace with rising sea-level is discussed in the next section, but for extensive areas of low-lying land it will obviously remain the only viable policy.

21. However, there are many areas where we should modify current coastal management policies to allow us to utilise the natural ability of salt-marshes to keep pace with rising sea-level. Just as warping may be used as a natural way of building up marshes prior to their final reclamation and draining, so natural marshes subject to regular tidal incursion will keep pace with sea-level rise provided the mud supply is adequate. As McCave (ref. 11) has shown in his study of the East Anglian coast (and others have shown elsewhere) mud supply comes from a combination of rivers and coastal cliffs. River supply will not increase with sea-level rise (though it is currently be well above natural levels as a result of agricultural land use), but the supply from cliffs will as we have seen. Even under current conditions, measurements have shown that coastal marshes are capable of adding 10 mm of sediment a year, so that as higher rates of sea-level rise will result in higher volumes of mud, marshes in favourable areas for mud supply will keep pace with sea-level throughout the next century.

22. In management terms this means (as we have so often seen) that if we can encourage natural changes, they will on

balance aid coastal stability even under conditions of rising sea-level. This is not to say that in some locations the outer edge of the marsh will not retreat, though this again will allow the mud to be transported to raise the general level of the remaining marsh. It also means that where marshes are narrow (naturally, or through reclamation), the whole system may have to roll landward by adding new flooded areas at the rear to match the roll-back of the marsh edge (or barrier beach) at the front. This envisages a transfer of land from agriculture to saltmarsh in the interests of maintaining natural (and thus cheap) stability, a technique neither provided for nor even envisaged in current legislation.

23. Rather less radical, though still reversing the trends of the past 150 years, would be the abandonment of flood-protected grazing marshes to natural saltmarsh and to natural accretion. It has become common to place floodbanks within natural marsh systems, with an outer zone left as saltmarsh and the inner zone turned over to grazing and/or the encouragement of freshwater conditions favouring certain bird species. As sea-level rises, these forward banks should be abandoned. Even under current conditions they have altered the stability of the outer marsh by restricting the tidal volumes flushed through the creeks, but as the reclaimed marshes lie increasingly below high tide level, the case for returning them to the naturally-accreting system will become increasingly strong. Interestingly, given current agricultural trends, it will probably be easier to return grazing marshes to the sea than bird reserves which rely on freshwater pools and associated freshwater habitats.

24. Inspection of the coastal map of the British Isles suggests that the intelligent management of coastal marshes to maximise natural adjustment to sea-level change is feasible along quite a high proportion of the low coast - indeed it will be an important method in helping to protect the flood banks which surround the more extensive areas reclaimed for arable farming and settlement. But quite a large part of the coastal saltmarshes have been reduced to a narrow belt which is neither wide enough, nor possesses an adequate system of creeks, to allow upward growth without undue loss of the seaward edge. In these areas only the early return of part of the reclaimed area to the saltmarsh system will allow upward growth to be resumed quickly enough to restore the coastal balance. If the areas are flooded too late, marsh levels may be too low, and the resulting higher transport energies may result in the loss of sediments and the formation of intertidal banks, not vegetated salt marshes capable of natural upward growth.

25. It may be worth noting that some of our natural coastlines may prove the most intractable. It may not be too expensive to induce farmers to abandon grazing marshes or return drained arable land to grass because the pressures of agricultural prices make these activities less profitable

than they were. But our scenario for both dunes and saltmarshes envisages a dynamic coastal zone, adjusting through natural processes as sea-level changes. New tidewater dominated habitats will be created on the landward margin as erosion allows the beaches and associated dunes to roll landward on the seaward side. Predominantly, current management policies for nature reserves and other sites of special scientific interest seek to maintain an existing range of habitats in their existing sites. Reports frequently welcome the addition of new land or shingle-ridge nesting sites to a reserve, but are consistent in deploring erosion elsewhere and hint that some action should be taken to reduce its impact. Our twenty-first century managers will have to accept that erosion is as natural a process as deposition, however unwelcome they find it, and that they must consider the overall inventory of habitats in the coastal margin, accepting losses in one place and gains in another. Indeed we shall need to promote natural change as a virtue and emphasise the scientific value of naturally evolving shores. At least the scenario outlined here suggests there will be more naturally evolving coast in the future as we utilise the natural capacity for adjustment which these zones provide.

THE THREAT OF INCREASED FREQUENCY OF TIDAL FLOODING ON LOW COASTS

26. Wherever extensive acreages have been protected from tidal flooding by relatively short sea defences, it will prove beneficial to maintain the existing conditions by spending money on raising and improving the floodbanks and reinforcing whatever protection they need from wave attack. Wherever possible, such defences should be located behind relatively natural outer defences such as wide beaches, coastal dune belts, and saltmarshes. In some cases retreat to an inner bank may be better than attempting to maintain a more seaward bank inadequately protected by a naturally-adjusting buffer zone.

27. Conversely, there will be other areas where the ratio of the area protected to the increased cost of flood protection will no longer show cost-benefit advantages, or where the flood banks are effectively already at their maximum economic height and where raising them would require major engineering works. It will also be necessary to consider whether the use of the coastal land as part of a naturally-adjusting coastal zone is better than its retention for agriculture. In these cases (and they are similar to the situation where a strategic retreat is made to an inner bank) we shall need legislation to allow compensation for the changed conditions. It will also be important to identify such areas at an early stage, so that scarce resources are not spent on banks soon to be abandoned, on drainage systems which will soon be redundant (or functioning as incipient marsh creeks), or wasted on improving existing property or

building new housing.

THE IMPLICATIONS OF THIS MANAGEMENT SCENARIO FOR LEGISLATION

28. Our coastal zone is a hazardous zone, managed even today by imappropriate techniques. The assumption has been that all hazards can be eliminated by engineering works, or in the case of the threat of North Sea surges, by a combination of flood defences and an untested warning system. Quite apart from the recent retreat from development control through planning by central government, planning has never been available to control building in this hazardous zone. On cliff-tops, planning permission is readily given for developments which are then defended by expensive engineering works - that the new or rehabilitated properties make the cost benefit case easier to establish adds insult to injury. In areas liable to coastal flood, Water Authorities are consulted, but their advice (e.g. to place a dormer window in a bungalow and to provide adequate access to a floored area of the roof space) is all too often ignored by the planning authority. Just as all cost-benefit cases on eroding coasts assume that the proposed works will completely stop erosion for periods of 60 years or more, so forward planning never anticipates changes despite evidence on maps and in the field that they are occurring.

29. If we are to adjust without financial stress to future sea-level rise we shall need to adopt planning measures that recognise coastal hazard and accept the reality of future coastal change. Either we require a change in planning controls and in policies on compensation in a coastal zone, or we need more general legislation which covers other areas of hazard such as river floodplains, methane-producing landfills, and such situations as landslide zones. The coastal or hazardous zone will require definition - we have seen that it will need to extend at least 1 km inland on eroding cliffed coasts and extend to the whole drained area in places liable to sea flood.

30. But above all, we need to understand that there are two golden rules governing effective and economical adjustment to sea-level rise. The first is to learn to use the natural ability of nature to adjust to changing conditions wherever possible, to allow cliff erosion so as to receive the downdrift benefits, and to encourage the natural build-up of sand dunes and especially salt marshes. The second and equally important rule is to decide what must be done at an early stage, for by adopting changed planning and compensation policies now, far less will be at risk by the middle of the next century and this will very greatly reduce the cost of adjustment. There is much to do and money will have to be spent improving sea defences in many areas, but if we avoid spending it in areas where other cheaper forms of adjustment are feasible, we shall keep the sea at bay without undue cost.

I should like to thank other members of the research team pursuing the impact of climatic change at the University of East Anglia under a contract for CEGB for discussion of the first draft of this paper. The views expressed are entirely my own.

REFERENCES

1. Bruun, P., 1988, The Bruun rule of erosion by sea-level rise: a discussion on large-scale two-and three-dimensional usages, *J. Coastal Research*, 4, 627-648.
2. MAFF, 1985, Financing and administration of land drainage, flood prevention and coast protection in England and Wales, (HMSO), *Cmnd*. 9449, 25 pp.
3. Bruun, P., 1986, Worldwide impact of sea level rise on shorelines, *Proc. Conference Climatic Changes, Washington, DC,: UN & EPA*, 4, 99-128.
4. Trenhaile, A.S., 1974, The geometry of shore platforms in England & Wales, *Trans Institute British Geogr*, 62, 129-142.
5. May, V.J., 1971, The retreat of chalk cliffs, *Geogr. J.*, 137, 203-206.
6. Cambers, G., 1976, Temporal scales in coastal erosion systems, *Trans Institute British Geogr.*, 1(2), 246-256.
7. Davies, J.L., 1980, *Geographical variation in coastal development*, (Longman, London & New York).
8. Pilkey, O.H. & Davis, T.W., 1987, An analysis of coastal recession models: North Carolina coast, in: *Sea-level fluctuation and coastal evolution*, ed. Nummedal, D., Pilkey, O.H. & Howard, J.D., (Soc. Econ Pal. & Mineral., Special Pub. 41), 59-68.
9. Dean, R.G. & Maurmeyer, E.M., 1983, Models for beach profile responses, in: *Handbook of coastal processes and erosion*, ed. P.D. Komar, (Chemical Rubber Company Press), 151-166.
10. Clayton, K.M. (in press), Sediment input from the Norfolk cliffs, eastern England - a century of coast protection and its effect, *J. Coastal Research*.
11. McCave, I.N., 1987, Fine sediment sources and sinks around the East Anglian coast (UK), *J. geol. Soc. London*, 144, 149-152.

15. Beach management

K. J. SHAVE, BSc(Eng), MICE, MIWEM, Technical Manager, Kent Division, Southern Water

SYNOPSIS. A wide variety of techniques have been used by Southern Water, and its predecessors, for the management of the sea defences along its coastline. This paper describes the selection and construction of a variety of new works from solid defences, using conventional concrete wave wall and apron, to flexible defences of shingle.

The maintenance aspects of the beaches and defences are reviewed historically to illustrate the development of the defences to the present day. This review indicates changes in policy that have been made over the years as management techniques have altered to accommodate revisions in the philosophy for new works which have been required to replace, or strengthen, defences that have reached the end of their effective working life.

INTRODUCTION
1. Historically many beaches, whether of sand or shingle, were backed by clay walls or large shingle banks, to provide the final defence against flooding. Over the centuries a combination of the relative lowering of the land in the south east of England in relation to sea level, a demand for ever improved standards of defence, and a depletion in the volume of the naturally accumulating shingle on the beaches as a result of coast protection works, has seen renewed major efforts in this century to secure the beaches, and strengthen the sea walls within the Authority's area.

2. These efforts began centuries ago on the south east coast and have continued up to the present day. Schemes have increased in size and value in real terms with the increase of the value of the areas protected against flooding, to the point where a scheme such as the one at Seaford can be promoted at a cost of £10 million.

3. This paper follows the historic development of shingle recharge schemes, pioneered by the Authority's predecessors over 30 years ago, in particular the first major scheme between Fairlight Cliffs and Winchelsea and the latest

Coastal management. Thomas Telford, London, 1989

at Seaford, the development of the Dymchurch sea wall and, as a contrast, the stabilisation works which have not only arrested the depletion of the dunes at Camber, but stabilised and increased them.

4. The failure of a variety of groyne systems over many years to influence beach profiles has resulted in the implementation of a policy which withdrew funds for groyne maintenance - with the result that beach levels recorded over a number of years have not significantly altered.

5. Since the Authority maintains sea defences only, and not coast protection works, this paper considers the bank or wall at the rear of the beach as an integral part of the beach in view of its potential influence on the mobile material that fronts it.

General Principles for Beach Management Options.

6. It is all too easy with hindsight to offer criticism of the way in which beach management has been conducted in the past, especially where it may have made a significant contribution to present day problems with sea defences. A multitude of authorities, each with their own particular problems, have historically tackled these as they perceived them, without always paying regard to the effects on adjoining lengths of coastline.

7. The advantages now of the Ministry of Agriculture's overview, and responsibilities in respect of grant aid for proposals for both coast protection and sea defence works, cannot be underestimated. This role, with its potential for moderation, when added to the Ministry's active involvement in promoting research projects and the setting up of data bases, and the work which CIRIA is undertaking on a national basis, leads to the conclusion that an overall management strategy is evolving for the coastline as a whole. In view of the input to these exercises from the widest possible experiences of the many contributors to these exercises, it could be that a series of "preferred solution options v problems" is ultimately produced that would relate capital and revenue expenditure to the new works options, and give an assessment of the likely overall success of the options for a solution to a beach management problem in the wider context of the effects on the coastline as a whole.

8. At present the general principles for beach management options would always begin with an assessment of the existing situation and a historic review of the beach and both past and current management practice.

9. A shingle beach, subject to annual recycling over many years, could easily come to the point where its

management is never questioned or reviewed and it is seen in the light of having only this one option for management. Clearly however, as modelling skills advance, it may be possible to combine to advantage recycling and a groyne construction/maintenance programme. Similarly, a solid wall, repaired and extended over many years may find itself locked into this cycle because the defence has not been breached or a collapse induced in the known past. These two perhaps extreme examples illustrate the likelihood that those responsible for beach management have often maintained the "status quo". In fact, the Authority has endeavoured to improve the performance of a shingle beach using groynes without gaining improvement, and has applied shingle recharge in front of failing solid defences with success. Nowadays, maintenance of the status quo is being questioned, although the "supporters" for example of groynes, or shingle recharge, or solid defences, or flexible revetments, still retain strong views as to their suitability or unsuitability.

10. The options for a sea defence appear to fall conveniently into two distinct parts - either a hard (inflexible) wall, or a soft (flexible) defence, both either with or without groynes. In practice, once the assessment stages referred to earlier have been completed, much recent experience has shown that designers have favoured a compromise of a solid wall fronted by shingle, which in many instances has been supported by groynes, such as the schemes at Minster for Swale Borough Council, Bexhill for Rother District Council, and Whitstable for Canterbury City Council. These are all coast protection schemes subject to widely differing wind/wave patterns and foreshore level, yet the same style of scheme has been designed. The Authority has adopted the same basic solution without groynes to sea defence problems at Pett, Walland, Littlestone and Seaford, except that the solid wall was in existence before the shingle was placed - in only one case, at Sheerness, neither groynes nor a solid backing defence were employed. Solid defences on their own will result in a stable foreshore sooner or later in many locations, and this is illustrated by the great wall at Dymchurch the development of which is described in the paper.

11. Offshore islands and rock headlands have been actively considered by the Authority for Seaford and the Northern Sea Wall at Reculver respectively - improvements in modelling and the results of full scale trials elsewhere may encourage the use of these options in the future.

Pett Sea Defences
12. Fairlight Cliffs to Winchelsea. Over the centuries, littoral drift, promoted from the west by prevailing south-westerly winds, maintained a shingle bank with an adequate height and bulk to form a natural sea defence.

COASTAL DEFENCE METHODS

The construction of the harbour arm at Hastings restricted, and then curtailed, further drift to the point where in the early 1930's the overall loss of material from the beach meant that protective works were required to prevent a breach.

13. Over a 3 year period 1933/36 the bank was protected along a 6.5 kilometre length at a cost of £200,000 with timber breastwork and a groyne system built to retard the movement of shingle. These measures were combined with a longitudinal wave screen of timber piles which was intended to reduce the wave impact on the wall and the beach immediately in front of it. Although extensive and costly at the time after only 10 years it had become clear that beach erosion was continuing and that the wall was deteriorating significantly.

14. In a further attempt to secure the wall another solid defence scheme was promoted at a cost of £700,000 over a length of 5 kilometres. This time interlocking concrete blocks were employed in a two stage wall incorporating an asphalt berm - further groynes were also installed with a view to arresting the drift of shingle along the foreshore. The toe piling became exposed very rapidly and it was clear that there was a real danger that the wall would be undermined and fail.

15. There were two options given active consideration :-
 (i) provide a "stronger" (inflexible) sea wall
or (ii) provide a shingle (flexible) defence.

16. The design produced for the first alternative would have extended and strengthened the existing wall, and taken the apron to a lower level - it would also have cost at least as much as the original wall. Further, there was no guarantee that the erosion of the beach would not continue to the point where this stronger wall would be undermined with the consequent ever continuing expense of maintenance and toe extension. The second alternative appeared to offer advantages - firstly, the beach levels were known to have fallen significantly since the advent of the solid defences and the loss of natural recharge material, and secondly, in the past, the shingle bank had provided a satisfactory defence and allowing for capitalisation of an estimated annual recharge, as well as the initial cost, this alternative was less expensive. The second option was adopted and 155,000 cu.m. of shingle were won from the accumulated deposits at the Rye Harbour western arm and placed in front of the solid defence in 1955. The western arm provides a very effective terminal groyne, as well as keeping the mouth of the river Rother clear, and the annual replenishment to maintain the shingle beach takes the material from this point and redistributes it at points along the beach.

17. It is interesting to note the parallel of events to part of the length of coastline between Broomhill (east of Camber) and Dungeness where depletion of a shingle bank was followed by reinforcing the clay wall with chestnut thatching and adding groynes to the foreshore. Refacing the wall at a later date with concrete blockwork encouraged further foreshore lowering until a shingle recharge scheme in 1957 arrested the decline. Similarly recycling of shingle maintains the defences along this length.

Dymchurch Sea Wall.
18. The Great Wall. This length of the Dymchurch sea wall from the northern end of Dymchurch northwards to the Grand Redoubt has a long history of beach management. A map of 1617 for example clearly shows references to groynes, some 'Y' shaped in plan and at that time there was probably a significant shingle beach. These groynes, or Knocks, made of stone, were the forerunners of modern rock/rubble groynes, since in 1847, the remains that were visible at low water, although separated from the line of the wall at that time, were seen "dividing at the end next the sea in the form of the letter Y". A map of 1837 indicates some irregular short groynes some of which exceeded 40 metres in length and some not connected to the main wall.

19. This effort at beach management had not been successful and as the beach became depleted with the accumulation at Dungeness and the loss of the naturally occurring feed., it was necessary to "arm" the clay walls behind the remaining shingle with "brushwood piles and overlaths". The walls were steep faced in some places, at 1½ or 2:1, and remained secure only so long as the shingle beach remained in front. As the shingle became depleted, this type of arming was continued down the face until the early 1800's when Kentish Ragstone facing was introduced and in the late 1830's, the slope of the face of the wall flattened to 6:1 and a curved wave wall introduced. Over the period to 1847 the length of the revetment had increased in some places and the ragstone was reinforced with rows of deal piling in the vicinity of the wave break for Spring Tides to limit the damage which could be caused by the waves. There is no record of groynes being part of these works and the engineer stated that the whole length of the wall was "in perfect state of repair" - by 1890 there was no shingle beach and the sand foreshore is reported as having fallen by 6 feet with the condition of the wall dangerous!

20. By the end of the century the practice of extending the revetment even further had been abandoned and Case had begun an extensive programme of groyning, apparently with some success, since sand levels were in places reported as having risen by up to 1.5m. It must be noted however

COASTAL DEFENCE METHODS

that the groynes were at intervals of between 35m and 60m and up to a maximum of 300m long, although on average about 100m long. Reconstruction of the groynes was necessary immediately before the 1939-45 war but on this occasion they were set to a pattern of 3 short (35m) and 1 long (75-120m) at 23m centres.

21. By 1959, despite maintenance, the efficiency of the system, except where the apron was to a flatter slope, was called into question as the groyne piles were now 5/6 feet above beach level and generally 1½/2 planks were exposed. The philosophy behind this latest arrangement of groynes can be questioned, but the foreshore had obviously lowered and damage to the wall was becoming more frequent and costly with a depth of water at High Spring Tides of about 6m.

22. It was concluded that the past practice of bedding ragstone on clay and grouting afterwards was no longer adequate, and a major reconstruction was undertaken in the early 1960's following model tests of the design profile by Hydraulics Research. The apron was extended seawards 6m and the toe protected by oak piling. The existing ragstone apron was grouted and the new one formed by placing 375mm cube concrete blocks set in mortar on top of mass concrete fill required to establish the slope of the apron.

23. Groyne maintenance was effectively stopped at that time and since the works were completed the foreshore level has remained stable.

Seaford Sea Defences.

24. The defences extend from the River Ouse at Newhaven some 4 Km eastwards and directly face the prevailing south westerly wind. The deep nearshore water allows for severe wave action on the defences.

26. The management of the beach has been significantly affected since before the end of the last century by the classic terminal groyne consequences of the efforts to keep the harbour entrance at Newhaven clear.

27. Early efforts, recorded from 1644, to keep the navigation open are successively recorded as providing only temporary relief. The natural replenishment of the Seaford frontage was not significantly affected until early in the nineteenth century when it was necessary to begin a programme of groyne construction in an area about a quarter the way along the beach.

28. The construction of the western breakwater between 1880/90 solved the problem of the navigation and stopped, at a stroke almost, the recharge of the beach by littoral

drift from the west. It also reversed the drift from west to east along the western part of the defences by altering the wave climate on this length and reversing the dominant angle of incidence.

29. Although acknowledged in a survey of 1877 by Captain Ardagh R.E., and in 1911 by the Royal Commission on Coastal Protection, that solid defences could cause erosion of the beach fronting the wall, the Board of Commissioners embarked on a programme of improving the solid defences in conjunction with groynes extending to low water.

30. Maintenance secured the defences until in 1936 possibly one of the earliest shingle recycling operations was undertaken to protect the wall. Shingle was transported from the eastern harbour arm, where it had been accreting as a result of the change in drift pattern, and re-distributed on the beach. This operation was supplemented by recharging at the same time with material brought from Dungeness.

31. During the periods of each of the World Wars maintenance was of necessity neglected and in places a fall in beach level of up to 3 m was experienced with low water mark advancing by some 60m. The consequences were increased wave action and ever increasing damage.

32. Additions to the solid wall were made at intervals over the years up to 1985. These increased the overall height of the wall as many of the works were to underpin, or reface, and prevent collapse, these being in addition to basic reconstruction of a total estimated length of 1.3 Km. During storms the mobile beach was disturbed to the point where wave action was direct upon the chalk seabed with consequent lowering of the bed.

33. Stability of the whole of the wall became of serious concern in 1981 when a 60m length subsided and tilted and again in 1985 when a major collapse was avoided by filling a void under the foundation of the wall with 240 cu.m. of concrete.

34. The Authority took over the Commissioners responsibilities in 1981 and in 1982 following a review of the history and condition of the defences began the process of promoting a major new scheme.

35. A major shingle feeding operation was seen as the most attractive of the options and Hydraulics Research undertook a study of this and also submerged breakwaters. The open beach solution was adopted and 1.45 million cu.m. of sea dredged shingle was imported and pumped ashore.

COASTAL DEFENCE METHODS

36. In order to secure the wall in the short term and provide against a length in the centre of the wall failing should a 5/6 day storm temporarily remove the recharge material from the face of the wall in that area, it was decided to place rock fill along the toe of the wall over this length.

37. In order to carry out the recycling necessary to maintain the profile a groyne was built at the eastern end of the wall to trap the shingle subject to littoral drift.

38. The new beach has now been subjected to severe storms and withstood them well. The shingle recycling pattern has yet to be established as drift has not all been in an easterly direction although actual losses of material from the beach are within the maximum calculated.

39. The improvement to the defences for the town of Sheerness and the surrounding area provides an opportunity to compare one part of the Sheerness scheme with the Seaford scheme.

Sheerness Sea Defences

40. At the Eastern end of these sea defences a 900m long naturally occurring shingle bank existed. This was maintained by material washed from the adjacent clay cliffs that naturally replenished the littoral drift along this shingle bank from east to west.

41. The bank was fronted by a very wide, almost flat, clay foreshore with a maximum depth of water of 4.5m at Spring Tides.

42. A major improvement to the shingle bank was undertaken in 1975 when 180,000 cu.m of shingle dredged from the North Sea, was brought to the shore at high tide by split bottom barges. After depositing the shingle, conventional earth moving plant formed it to the correct profile.

43. It was acknowledged that natural replenishment would not be sufficient to maintain the profile. This part of the scheme has a "terminal groyne" (an old coal jetty) where it was anticipated that the bulk of the littoral drift would accumulate for recycling and this has proved to be the case.

44. No groynes were included in the design since they would have been large and expensive to construct, and their need doubtful to maintain the beach profile. Also much of the other lengths of the defences to the west rely on shingle to protect the toe.

45. The specification for the shingle limited maximum size and attempted to match the grading curve to that of the existing shingle. Also, the maximum permissable percentage of fine material below 6mm was stated as to be "not greater than would fill 50% or less of the voids in the larger material" - if this could not be obtained then mixing from different sources was required. At Seaford it was acknowledged that the far greater quantity of shingle employed, almost ten times the Sheerness volume at 1.45 million cu.m, a different approach could be followed with a more relaxed specification where up to 25% by weight of material passing the 5mm sieve could be placed.

46. At Sheerness any significant "washing out" of fines by wave/storm action that caused a depletion in the bulk of the bank would have had significant consequences for the ability of the bank to meet its design criteria. At Seaford the bulk of the beach is such that the design assumes loss caused by sorting to a depth of about 2m without affecting the performance of the beach, and it was predicted by Hydraulics Research that possibly 200,000 cu.m could be lost in this manner.

Camber
47. As the Camber sands increased in popularity and use as a holiday resort pedestrian traffic across them damaged the natural vegetation to the point where it could no longer effectively prevent wind depletion and erosion of the dunes. Not only did this potentially have an effect from the point of view of the sea defences but also caused severe problems for the community sheltered behind the dunes. The increase in wind blown sand caused roads to be covered and following entry to roof cavities, ceilings to collapse.

48. A jointly funded programme of stabilisation was undertaken by Rother District Council and the Authority's predecessors commencing in 1964 and this is continued annually.

49. The programme of promoting build up of sand using permeable screens of brushwood or wattle fencing followed by planting of marran grass has proved entirely successful as have experiments with grass seed. Any erosion at the seaward toe of the dunes is quickly made up by wind blown sand from the wide foreshore fronting them.

50. Similar work has been successful at Greatstone.

Conclusion
51. Although the Authority maintains 208Km. of sea walls in Kent, much of these sea walls are in estuarial conditions, fronted by mud, exposed only at low tide and

not subject to severe wave action. Where beaches front these walls the paper has highlighted the change of philosophy in management of the beaches from solid defences of concrete or slabbing or blockwork to the provision of flexible shingle beaches maintained by recycling.

52. The change in philosophy does not however extend to the point where amenity considerations are ignored. A strong argument has been considered in the Dymchurch area for the protection of the Great Wall by shingle recharge but in view of the stable levels of the sand foreshore has been rejected in favour of significant capital expenditure refurbishing the existing wall.

Acknowledgements

The author wishes to thank Mr. B.A.O. Hewett, Managing Director (Water Services) Southern Water for permission to publish this paper, and to Mr. A.E. Holmes former Engineering Manager of the Sussex Division, for allowing his paper to the 1988 IWEM Conference on the Seaford Sea Defence Scheme to be used as a basis for part of this paper. The authors views expressed in the paper are not necessarily those of Southern Water.

16. Marine resources

F. PARRISH, Head of Foreshore and Seabed Branch, The Crown Estate

SYNOPSIS. When considering coastal works, but particularly those involving each recharge, it is natural to think in terms of uses of seabed materials. Environmental and practical considerations suggest that this will invariably be a better solution than the use of materials brought from land sites. This paper seeks to give a perspective on the use of such material from the point of view of the resource itself and to explain the procedures which, in the absence of a statutory planning framework, have been adopted for considering the release of areas of seabed for resource dredgings.

THE CROWN ESTATE
1. The Crown Estate is not a Government Department in the commonly understood sense. Under an Act of Parliament (The Crown Estate Act 1961) The Crown Estate Commissioners are constituted as a body corporate charged with managing and turning to account land and land rights held in right of the Crown. The Commissioners' general duty is to the Estate itself - to maintain and enhance it and to obtain the proper return from it, but "with due regard to the requirements of good management". Under historic arrangements the Estate itself is owned by the Monarch "in right of the Crown" but since 1760 the annual revenues have been passed to Parliament and in return Parliament undertakes those costs of government which previously fell on the Crown's income and grant the Civil List payment.

2. The foreshore and bed of the sea is a very ancient Crown possession. The Crown Estate still owns roughly 55% of the foreshore - the land between mean high and mean low water, most of the bed of the territorial sea (now 12 miles), with very few exceptions, and beyond the territorial sea limit it owns the rights to explore and exploit the natural resources of the continental shelf. In all cases of course oil, gas and coal are excluded as they are from other estates both on land and water.

DEMAND FOR SEABED MATERIAL
3. Leaving aside special requirements such as the small amount of dredging for tin, waste coal and lithothamnion the main demand

Coastal management. Thomas Telford, London, 1989

at present for seabed minerals is for sand and gravel. By far the most important demand is that for aggregates for concreting and building use. From very small beginnings the industry has grown, most spectacularly since the early 1960's, to provide 15-16% of the UK's total requirement for building aggregates. In some counties, notably Hampshire and Sussex, the marine proportion of the market is very much higher, recently around 50%. The demand for marine aggregate for this market is strong. Most of the arrangements which are in place for licensing the use of seabed materials have grown from the requirements of the main marine aggregate producers. This demand is mainly for a good balanced mix of sand and gravel - ideally in a concreting proportion of 60/40 although the days when this mix could be won in situ from the seabed now seem generally to be past. About 18 million tons was dredged for this purpose in 1987, nearly 11% up on the previous year.

4. The second main requirement is for material to be used for land fill for land conditioning or for coastal works. For example, infill for Docklands in London, Cardiff and Southampton, reclamation of land for port works at Dover and Portsmouth and material for producing better quality land from marshland in the Thames estuary have all been markets for marine material in the very recent past. In 1987 nearly 2.5 million tonnes was won directly to meet this requirement and there was an additional unquantifiable amount which was landed through the normal ports of landing but used for the same purpose.

5. The final main requirement and that which most concerns the present readers is that for beach recharge and replenishment. In 1987 material won directly for this purpose was over 3 million tonnes - the vast majority of this being for Southern Water Authority's projects at Seaford and Glynde Gap.

6. Over the past few years there has been a steady demand averaging between 2 and 3 million tons each year for coastal work whether for coastal protection, beach replenishment or land fill. All the indications are that this demand also will continue to increase and we are already aware of several major projects which will create very substantial demands for seabed material above and beyond the normal aggregate market demand if they come to fruition. The type of material required varies from sand to something approaching an aggregate specification.

LICENSING

7. Against the background above it is important to understand the basis of the licensing procedure, the constraints on release of areas and the competing demands for seabed use. Basic procedures are very much the same whether the end use is for aggregate purposes or for coastal work. The basic type of material in demand also is, as explained above, very similar - in effect only the grading curve differs according to the requirement.

8. The present arrangements date in essence from the early 1970's. The Commissioners require that before considering an application to extract sand and gravel from the seabed the prospective applicant must satisfy them that it has the resources and expertise to meet the licensing conditions and that the applicant should take a prospecting licence.

9. Prospecting licences cover defined areas of seabed, and permit the holder to carry out prospecting by seismic survey, sidescan sonar and grab samples. In addition the licence generally permits a limited quantity to be taken by dredger sampling.

10. Once a resource has been identified and prospecting carried out in accordance with the terms of a prospecting licence, the applicant may apply for a production licence. There is no statutory basis, beyond the powers of the Department of Transport under the Coast Protection Act which relate solely to navigational interference, under which an application can be assessed. What has therefore evolved is an informal procedure under which the Commissioners have agreed that they will only grant licences if, following consultation with the relevant Government Departments, there is no substantive objection to the proposal.

11. Our first step in considering an application is to refer the proposal to Hydraulics Research Ltd (HRL), an expert consultancy, to advise whether there is any likelihood of the dredging causing damage to the adjacent coastlines. The information given to them is a chart showing, by co-ordinates, the proposed area of extraction, the prospecting report, and a proposed maximum extraction, either annual (normally based on 5% of the likely reserves) or total.

12. In forming an opinion on the licence application HRL aims to answer the following questions:

 12.1 Is the area of dredging far enough offshore so that material is not drawn down the beach into the deepened area?

 12.2 Is the areas to be dredged sufficiently far offshore and in deep enough water so that changes in the wave refraction pattern do not take place? Such changes may alter the longshore transport of beach material and hence affect shoreline stability.

 12.3 Does the area of dredging include offshore bars which are sufficiently high to give protection to the coastline from wave attack? A significant reduction in the crest height might increase wave action at the shoreline and lead to erosion.

12.4 Is the dredging to be carried out in deep enough water so that it will not affect possible onshore movement of shingle?

13. After an initial assessment of the proposal, HRL informs the Crown Estate of the extent of research they consider necessary. This may include desk study, computer simulation, site inspection or in some cases a full-scale field research programme. The Crown Estate requires the costs of this study to be met by the applicant who must decide on the basis of the information provided by HRL whether the cost of the research required is economically justified. If the applicant feels unable to meet the costs of the research considered necessary the application proceeds no further.

14. In consulting HRL the Crown Estate seeks an assurance that the dredging proposed will have no significant effect on the coastline - the 'zero effect principle'. Given that we are dealing with a low cost commodity and a high level of uncertainty whether any production application will be granted, it is well understood by the Crown Estate and HRL that what is required is the highest measure of surety at the lowest economical price. The result of these requirements is that HRL's judgements err well on the side of caution. The computer model itself tends to over-estimate effects and the requirement of zero effect means that the consultant's views will be based on the worst case. If, on this basis, HRL's view is that there would be deleterious effects on the coastline the company is informed, and the application goes no further. Before reaching this stage, however, it is common for HRL to discuss its potential results with the Crown Estate and its advisers and with the applicant to see whether by placing restrictions on the proposed dredging (e.g. a depth limitation on the lowering of the seabed allowed) or by sponsoring further and more expensive research, the difficulties might be overcome.

15. Presuming that HRL offer a favourable view and that our checks on existing pipelines, cables, etc. have shown no obstacles, the application, supported by HRL's report, the prospecting report, and the Crown Estate's own views is put to the Department of the Environment (DOE) Minerals Division for a 'Government View'.

16. Minerals Division obtain this view by consulting all Departments whose interests might be affected by the proposal. It is, in effect, a 'cascade' consultation procedure under which Minerals Division will go initially to:

16.1 Other Divisions within the Department of the Environment (Construction Industries Division, the sponsor for the industry, Rural Affairs Division on Environmental Aspects).

16.2 Ministry of Agriculture, Fisheries and Food, (Fisheries and Coast Protection).

16.3 Department of Transport (Navigation) who will indicate at this stage whether they will be able to issue a consent under the Coast Protection Act.

16.4 Ministry of Defence (hydrographic and Naval interests).

16.5 Department of Energy (Oil etc. interests).

16.6 Riparian Councils

17. Each of these Departments will in turn consult other bodies such as local coast protection authorities, fishery committees, Nature Conservancy Council, regional water authorities, ports, etc. Some of these bodies will consult further.

18. The Government View effectively rests upon whether or not any Government Department sustains an objection to the proposal. As soon as a substantive objection to the proposal has been raised Minerals Division will inform the Crown Estate who, after discussion with the applicant, see whether it is possible by negotiation, by further research, or by changes in the area or licence conditions to resolve the objection. Thus although the initial consultation period is limited to a matter of a few months, the final resolution of an application has in the past sometimes taken several years. If it is not possible to resolve any objection then a favourable view is not given and a licence will not be issued.

19. It may be worthwhile to give some examples of the kind of objections which have been raised and the ways in which compromise solutions have been sought and found. The most common grounds for objections, and the ones where there is most scope for compromise, are coast protection, fisheries and navigation.

COAST PROTECTION

20. The HRL report which accompanies every application seeking a Government View forms the basis for discussion of coast protection issues. HRL, once they have given a favourable view in their report, will act as independent consultants in considering any further objections. HRL's research has shown that, as a general rule of thumb, dredging below the 18-metre depth contour does not produce discernible effects on nearby coastlines. Even then however objections can sometimes lead to modification of an application, for example, by restriction on the dredging depth permitted or the excluding crest areas which might afford some protection for inshore spending banks. In other cases it is possible by further research to demonstrate that dredging can safely be permitted inshore of the 18-metre contour, for example, subject to annual bathometric surveys and

further monitoring by HRL of the progress and effects of dredging. It should be noted that although the hydrodynamic studies are carried out economically they are nevertheless based upon HRL's considerable expertise and experience in this field and carry the full weight of their national and international reputation.

NAVIGATION

21. Generally navigational aspects are fairly straightforward. Initial objections can sometimes be raised as a result of misunderstanding about the manoeuvrability of dredgers and these can be dealt with comparatively easily. In some cases a specific requirement can be placed in the licence stipulating that only trailer dredging may be carried out and that no dredging will be done whilst at anchor. In other situations the reverse might apply. A common requirement is that dredging will take place in line with the general flow of shipping traffic in an area. Apart from direct interference with navigation, there might be fears about the dredging affecting existing navigational channels. Where necessary further studies by HRL can validate or remove such objections.

FISHERIES AND ENVIRONMENT

22. The inter-relationship between fishing and dredging is probably the one which causes greatest difficulty. It demands very close consultation between the Crown Estate and MAFF on all dredging proposals. In addition to the Government View procedure therefore a Code of Practice was agreed with MAFF, the terms of which were published in December 1981. The aim stated in this document is

> "To provide a basis for close liaison at working level between the fishing and dredging industries in order to promote mutual co-operation and to reduce to a minimum potential interference with each other's activities and damage to each other's resources".

Before a prospecting licence is issued, the Crown Estate informs MAFF headquarters of its intention to do so and MAFF consults its Fishery Laboratories and the local District Inspector of Fisheries. They then inform the Crown Estate and the company whether there are likely to be potential conflicts in the area in question. Once a prospecting licence has been issued the Crown Estate sends details to MAFF Headquarters and the District Inspector and at that stage the local Sea Fisheries Committee is also informed. The applicant is also put into contact with the District Inspector and the Sea Fisheries Committee so that right from the start working contacts can be established and problems discussed. In particular any proposals for dredged samples require clearance by the Crown Estate and in each case MAFF is consulted on the timing and the area concerned. In this way effects of sampling on, for example, spawning grounds or shellfish beds are minimised.

23. At the production application stage, under the Government View procedure, MAFF, as explained above, consults its laboratories, the District Inspector and the Sea Fisheries Committee which co-ordinates local fishery interests. Under the Code of Practice, where MAFF intends to object to a licence it notifies the Crown Estate and the applicant, explaining its concerns. Invariably, these prove both contentious and time-consuming if they can be resolved at all.

24. Many difficulties arise from the paucity of information on a local scale on the volume, type and quality of fishing in particular areas. By the time a dredging application gets to a production application stage we are talking typically of an area of less than 50 square kilometres, in many cases less than 20 square kilometres; ICES squares on which fishery information is generally available are typically 1,100 square kilometres in extent. The interactions between dredging and the environment are complex; in general terms however it is well accepted that while there will be some damage to commercial fish stocks from the dredging operation itself, these are small and except in unusual circumstances insignificant.

25. Each application is therefore considered on its own merits, taking into account the extent and type of benthos, the possibility of change in the seabed environment, the existence of spawning grounds for herring and sand eel, shellfish beds and fixed gear fisheries.

26. In a paper of this scope it is impossible to cover the range of interactions or the scientific studies which have been carried out. Unless however MAFF can be satisfied that the proposed dredging will not have a significant effect on commercially important fish stocks or their food chain, a favourable Government View will not be granted.

27. The question of environmental effects of dredging is often confused and overlaid with passionate arguments about physical interference between dredgers and fishing vessels and gear. Clearly, in considering whether to raise an objection MAFF has regard to the size and the commercial value of the fishery concerned. Without doubt, some fishery activity will be affected by the dredging activity. In general however with good co-operation and prior information in accordance with the Code of Practice the disturbance can be minimised and the cases of direct conflict within established dredging areas are very few indeed. Trawlers can work quite successfully in the close vicinity of dredgers. Fixed gear fishermen clearly cannot work in an area which is subject to regular dredging but again the Code of Practice is designed to ensure that information on dredging activity within a particular area is passed regularly to the local District Inspector and Sea Fisheries Committee so that areas which are not actively being dredged can be used by fishermen.

28. It is important to remember too that while dredging grounds may look extensive on Admiralty charts, the actual area being dredged at any one time is comparatively small. For example, if one assumes an average dredging depth of 1m, the actual area of seabed actually touched by dredging pipes in a year amounts to about 9 square kilometres, a tiny proportion of the total UK Territorial Sea and Continental Shelf. Within this of course there is considerable scope for intense localised conflict but it is our firm view that given good communication between dredging and fishing industries at all levels, such conflicts can be resolved, or at least ameliorated.

PRODUCTION LICENCES
29. Only when all objections have been resolved and a favourable Government View has been issued is a production licence issued. The licence from the Crown Estate stipulates, by reference to an extract from the Admiralty chart and co-ordinates, the area in which it is permitted to dredge: it lays down either total or annual permitted maximum and it also lays down any conditions required by the Government View itself. For example, in a recent case the Government View conditions required that the area could only be dredged by suction trailer dredgers; that the area should be buoyed; that only those areas within the licensed area where it had been proven that a sand and gravel thickness of over 1m in depth existed could be dredged; that a capping layer of sandy gravelly material of not less than 0.45m was to remain after dredging; and that in no circumstances was the overall depth of the dredging to exceed 2m. In other cases, there have been requirements for seasonal adjustment, or night dredging only.

The licence also stipulates the annual royalty rate. Royalties are payable per tonne of material extracted from the seabed.

RECENT CHANGES
30. The EC Directive on Environmental Assessment covers the offshore area as well as land. Under the Directive any application for a mineral activity which is likely to have a significant effect on the environment will have to be accompanied by an environmental impact assessment. A consultation paper by the DOE on the implementation of the Directive proposes in essence that this requirement will be incorporated into the Government View procedure. The DOE's consequent revisions to the Government View procedure itself are awaited.

31. Criticisms of the present system which have been raised by the dredging industry are:

> 31.1 There is no single Government authority responsible for developments in Territorial Waters and the UK Continental Shelf. Many Departments and organisations have interests in the seabed and separately keep information relevant to their

interests. This may lead to unjustified refusals or unnecessary restrictions being placed on dredging.

31.2 There is no overriding authority or mechanism within Government with responsibility for deciding a particular application on its merits. If one Department maintains an objection this is sufficient for an application to be refused whatever the merits of the case.

31.3 The present arrangement does not allow dredging companies sufficient opportunity to explain their case to potential objectors in public.

31.4 Applications for licences take too long to determine.

In addition to the concerns of the dredging companies there have also been criticisms from environmental bodies, recreational associations, local pressure groups and the general public who do not have a formal opportunity to comment on proposals under the existing system. In extreme cases local people may be unaware of dredging proposals because there is no requirement to publicise applications. As a result the DOE has also considered the whole of the Government View procedure, and has taken in several proposals from the Crown Estate.

32. The Crown Estate will in future take a more positive role in assessing proposals prior to submitting them to the Department of the Environment for a formal Government View. We have recently employed an environmental scientist who is charged with carrying out early informal discussions on dredging proposals with the companies, and with agencies, bodies and Government Departments whose interests might be affected. On the basis of these consultations a full report on the possible effects of dredging proposals and the possibilities for overcoming or ameliorating them will be prepared. In this way an application for a Government View will in future be accompanied by a full statement setting out the pros and cons of the application in a considered manner thus allowing the Government Department concerned to reach a formal Government View more expeditiously. A fundamental part of these arrangements will be local advertisement of the licence application with sufficient detail being made available to allow sensible comment on it. We hope that in this way many of the criticisms levelled at the present procedures can be met while still preserving the essential requirement that the Government should take a formal view on the desirability of any dredging proposal.

CONCLUSION

33. I hope it is clear from what I have said about seabed material, far from being a valueless and limitless commodity which is there for the taking is in fact a product which is in considerable, continuing, and growing demand; which competes for

COASTAL DEFENCE METHODS

sea space with a wide range of sensitive and established alternative uses and which can only be released after thorough planning and consideration. It is important to take into account the timescale for the release of any substantial quantity of material for use. The Crown Estate charges a royalty on material used and this royalty will take into account all normal factors which go into a market price, most notably, competing demands, the going rate for similar material, the quality of material and in some cases the use to which the material will be put. We and our consultants, Posford Duvivier are always ready to advise on the implications of any project even at the early planning stages; all too often we are approached at a late stage of planning, sometimes even after a contract has been let, and whilst, to the best of my knowledge and belief we have never delayed or failed to meet a real requirement it has sometimes been a very close-run affair. If I have done no more than get this message across to any potential user then I will consider my time and at least some of yours well spent.

17. Management of coastal cliffs

A. McGOWN and L. K. R. WOODROW, University of Strathclyde

SYNOPSIS
There is a history of coastal erosion along the south and east coasts of England associated with the cliffs formed of Tertiary and Quaternary deposits. A review is given of the coastal processes affecting the coast and in particular the geotechnical factors affecting the stability of clay and soft rock coastal cliffs. Two case histories are presented to illustrate typical coastal cliff stability problems which highlight the need for greater consideration to be given to the stability of eroding and protected coastal cliffs. A discussion is also presented on the management of coastal cliffs in the context of the Coast Protection Act (1949).

INTRODUCTION
1. In the United Kingdom the problem of coastal erosion is most severe along the south and east coasts of England where the cliffs are composed of Tertiary and Quaternary soils and soft rocks. The resulting erosion and flooding along this coastline has resulted in a considerable loss of land and large expenditure on coast protection measures.
2. The Department of Environment has reported that, "at present, virtually all the urban areas on the coastline which require or merit coast protection are either protected or will be protected in the near future", (Herlihy, 1982). Nevertheless, the problem of coastal erosion has not been entirely overcome as there remains the problems of maintaining and replacing existing works and of erecting new works to protect recently developed land. Also, there is a particular problem relating to maintaining the stability of coastal cliffs formed of soils and soft rocks. In the past, the stability of these cliffs was considered to be of secondary importance compared with the prevention of beach erosion and was in any case thought to be dealt with by the construction of toe erosion protection works. However, coastal cliff failures may occur some tens of years after the provision of protection works, with resulting damage to property and distortion or sweeping away of the works at their bases. Therefore, in order to manage effectively coastal areas it is essential that the long term behaviour of both protected and unprotected cliffs be understood and measures to deal with them incorporated into a Coastal Management Scheme.
3. This paper reviews the coastal processes affecting the coast and in particular the geotechnical factors affecting the stability of coastal cliffs. Two case histories are presented to illustrate typical coastal cliff stability problems. A discussion is also presented on the management of coastal cliffs

COASTAL DEFENCE METHODS

in the context of the Coast Protection Act (1949).

COASTAL PROCESSES
4.   The coastline is the interface between the terrestrial and maritime environments, therefore the morphology of the coasts is governed by both.
5.   The rate of coastal erosion is dependant upon the relation between the assailing wave forces at the cliff base and the shear strength of the cliff forming materials. The morphological process tends to be cyclical and can be summarised as follows:

(a)   Erosion by the sea of the toe of the coastal cliff with transportation of debris alongshore and offshore.
(b)   Oversteepening of the cliff face.
(c)   Failure of the cliff.
(d)   Deposition of material at the toe of the cliff.

6.   When waves erode the base of a cliff, instability is induced due to an increase in slope angle which induces increased shear stresses within the slope. Slip debris is accumulated at the base of a cliff and protects the cliff base from wave attack. Continuing wave action erodes and transports the debris either alongshore or offshore, which again exposes the cliff base to further erosion. The longshore wave activity controls the movement of sediment along the coast, which accounts for zones of coastal accretion and erosion, and the development of various landforms such as spits, barriers, salt marshes and other features.
7.   The height of the cliff and rate of removal of material from its base influences the mechanism of slope failure. A low rate of removal allows the slope to degrade naturally and the processes of swelling, softening and weathering begin to contribute to the causes of instability. The mechanism of failure may be a combination of rotational or translational slides and mud flows. The contrary case is where removal of material is rapid and the base of the cliff is under continuous attack. During this condition erosion is rapid, with falls and topples being the primary mechanism of failure.
8.   The maintenance of this complex process of erosion and accretion is an important element of coastal management. Removal of sand and gravel for the construction industry from beaches and near-shore areas around the British coast has largely ceased in areas identified as being prone to erosion, but there are numerous historical cases where severe erosion has followed the removal of these materials. These case histories are a reminder of the importance of beaches in coast protection.
9.   The removal of sand and gravel from the offshore zone can also raise problems with regard to coastal erosion. A system of licensing by the Crown Estate Commissioners for offshore dredging is currently enforced in the United Kingdom. However, dredging for navigational purposes by Port Authorities is outside the licensing system. In all cases, Coast Protection Authorities should be concerned about the removal of marine deposits and its possible consequences on coast erosion, since the removal of large quantities of offshore sands and gravels may alter the offshore and inshore hydraulic regime and so the stability of the beach and inshore deposits.
10.   The lack of natural beaches is a fundamental cause of severe erosion and damage to coast protection works. Various methods of maintaining an adequate beach defence have been used, including the construction of groynes and artificial beach feeding. However, the effectiveness of some beach retention

schemes means that the overall coastal system of which they form a part, is losing a large proportion of the natural littoral drift. This can affect the formation of natural beaches elsewhere in the system.

GEOTECHNICAL PROCESSES
11. Toe erosion by the sea is often the main causal factor that promotes landsliding in unprotected or partially protected clay coastal cliffs. Other factors also influence the stability of the cliffs and may become dominant once erosion has been prevented. Thus, it is important to recognize that the protection of the slope from erosion will not necessarily ensure the long term stability of a slope. A protected, ("abandoned"), coastal slope will continue to degrade until it reaches a stable angle of repose. The mechanism of degradation may involve deep seated movements as well as shallow surface slides, mud flows and soil creep. Under natural conditions this may take hundreds, if not thousands, of years to complete and involve substantial loss of cliff-top land. For example, an investigation by Hutchinson et al (1976) of an abandoned London Clay coastal slope at Hadleigh, indicates that it will take in excess of 10,000 years for the slope to reach its angle of ultimate stability.
12. The degradation of slopes usually occurs in an episodic manner due to weathering, changes in pore water pressures and climatic events. This process may effect the stability of building developments on or near the slope and coast protection works at the toe of the slope.
13. A delay prior to slope failure, is a common phenomenon that has been observed both in excavated slopes and protected coastal slopes. Failure occurs as a consequence of the rate of equilibration of pore water pressures developed post-excavation or post failure within the soils and soft rocks forming the slope. Unloading caused by excavation or failure causes a reduction in pore pressures in the soil around the failed or excavated area. With time, these pore pressures will tend to equilibrate and a steady state long term pore pressure regime will be reached. The time for depressed pore pressures to equilibrate depends on the permeability, the drainage path length and the swelling characteristics of the soil. Chandler (1984) reported on a number of cases of observed pore pressure equilibration and suggested that they fell into two categories, as summarised below :

(a) Normally or lightly over-consolidated Quaternary clays which are either strongly fissured or else have fairly frequent permeable horizons, and have in-situ permeabilities in the range 6×10^{-6} to 1×10^{-9} m/sec: Slopes achieve long term pore pressures either during excavation or within a few months of excavation being complete.
(b) Heavily over-consolidated clays or clay shales, again fissured, but with in situ permeabilities often less than 2×10^{-10} m/sec, and with few permeable horizons: The attainment of long term equilibrium can be a lengthy process, varying from a few years to many decades.

14. The above work reported by Chandler is supported by case studies on the observed long term pore water pressure increases in London Clay coastal slope at Herne Bay, North Kent by McGown et al (1987). This work reported rises in pore pressures in protected coastal slopes over a period of 20 years. A historical survey indicated that major deep seated failures occurred at this site on a cycle of between 50 to 80 years.

From the back-analysis of these failures it was suggested that the slopes failed before complete equilibration of pore pressures had been achieved.

15. Thus the work of Chandler (1984) and McGown et al (1987), amongst others, gives a clear indication of the possibility of delayed failure within slopes and highlights the need for long term pore pressure monitoring in slopes in order to assess their long-term stability.

16. Groundwater and the change in groundwater flow are also factors that have a major influence on the stability of slopes. The effective control of groundwater using drainage is an important element in improving the stability of a slope, whereas poor control on groundwater can cause instability. The principal adverse actions influencing the groundwater regime and their effects on coastal slopes are outlined in Table 1. The "natural actions" are sometimes difficult to control, especially on eroding coastal slopes, but not so the "man-made actions". Drainage from agriculture, gardens, highway and domestic drainage can all contribute to slope instability if outfalls are not properly designed and controlled.

Table 1. The Effects of Groundwater on the Stability of Coastal Slopes.

ACTIONS

Natural Actions
    Erosion
    Landsliding
    Debris accumulation
    Sea level fluctuations
    Meteoric events
    Earthquakes

Man-made Actions
    Cutting and filling
    Groundwater changes - stabilisation
                              - agriculture
                                - domestic
                                - highway etc.
    Removal of vegetation, (tree and shrub cover)
    Irrigation
    Construction on slopes
    Construction near the top of slopes
    Consequential effects from other areas

EFFECTS

Processes	Consequences
Pore pressure increases	Reduction in shearing resistance
Internal erosion	Changes in permeability and compressibility
Softening	Reduction in shear strength parameters
Weathering	Modification of the physio-chemical characteristics and reduction in shear strength

GEOTECHNICAL ASSESSMENT

17. Assessment of the stability of natural slopes can pose many problems. Slope stability predictions are dependant on the accuracy of the method used to model failure and the input parameters, such as soil strength and pore pressures. Pore pressures can be measured in-situ, but the determination of suitable soil strength parameters to incorporate into the analysis is more difficult. Case studies comparing the strength of soils obtained from laboratory testing compared to soil strength determined from back-analysis of failures often indicate marked differences. Reasons for this, based on data from London Clay failures, indicate the mechanism involved is progressive, i.e. failure does not occur simultaneously along the entire slip surface, and the mobilised strength at failure generally lies between the peak and residual values. However, observations on glacial till of lower plasticity than London Clay indicate strengths mobilised at failure close to laboratory peak values, Skempton et al (1961). Minor contributions to the mechanism of failure are also made by softening and rheological effects.

18. The above discussion illustrates the need for a complete understanding of the behaviour of soils and soft rocks forming a cliff, and of the mechanism of failure, before reasonable predictions can be made of the stability of a slope. A useful technique is to base predictions on the results of careful back-analysis of previous failures at the same site.

CASE HISTORIES

19. The following case studies illustrate two typical problems associated with coastal cliffs encountered on the south and east coasts of England.

Scarborough

20. The effect of the growth of a coastal town is well demonstrated by the development of Scarborough on the North Yorkshire coast. In the eighteenth century, the town was built on the relatively stable ground to the lee of the Scarborough Castle headland. In the nineteenth century, the town rapidly expanded toward the north and south along the tops of eroding coastal cliffs. The coastal slopes were partially stabilised by shallow surface drainage and laid out as public gardens. Despite the construction of adequate coast protection in the form of a sea wall during the later part of the nineteenth century, the slopes continued to fail. Indeed, The Scarborough Gazette in 1896 reported, "One of the unpleasant and expensive features of the foreshore at Scarborough is its liability to landslip. Even where the cliff is protected by sea walls this evil cannot be wholly avoided,...", Anon (1896). Figure 1 shows the North Bay of Scarborough, circa 1890. A problem now faced by the Local Authority is the stabilisation of their urban coastal slopes in order to prevent failures affecting development at the top, face and toe of the slope.

21. Figure 2 illustrates a landslide in the North Bay area of Scarborough. The slide is shallow rotational, in glacial detritus and extends from the toe of a 25 metre high, 1:3 to 1:4 slope. The cafe in the foreground of Fig. 2 has been cut into the base of the slope. Water flows from the slide area, probably originating from broken surface water drains. It can be suggested that failure is probably due to the deterioration in performance of slope drainage in an oversteepened slope. This type of problem is more acute for the Local Authority because the instability is not directly attributable to erosion by the sea, therefore it would appear that application cannot

COASTAL DEFENCE METHODS

be made under the Coast Protection Act (1949) for an Exchequer Grant. This is not an isolated problem but one which is fairly common in seaside resort towns around the south and east coasts of England and is likely to become more apparent as existing slope stabilisation works deteriorate with age.

Figure 1. North Bay Scarborough, circa 1890.

Figure 2. Landslide, North Bay, Scarborough, 1988.

Christchurch Bay
22. The Christchurch Bay area illustrates a coastal problem where a number of maritime authorities share responsibility for a short length of coastline. In this case responsibility is shared between the Bournemouth, Christchurch and New Forest

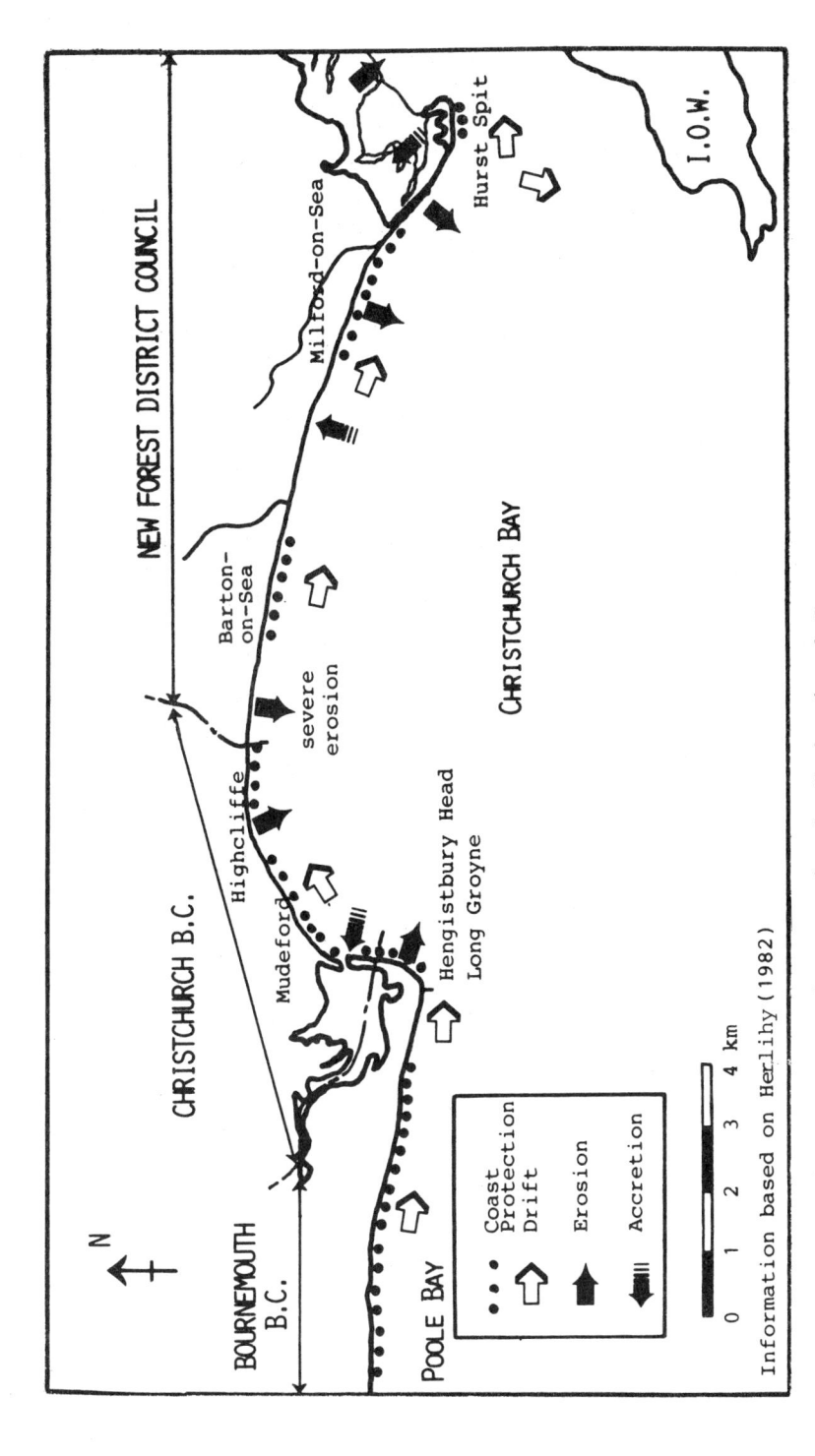

Figure 3. Coastal processes and protection in Christchurch Bay

COASTAL DEFENCE METHODS

Local Authorities, Fig. 3.

23. Christchurch Bay forms a natural coastal unit with its boundaries at Hengistbury Head and Hurst Spit. The cliffs range in height from about 30 metres at Hengistbury Head and Highcliffe to about 10 metres at east of Milford on Sea. Between Hengistbury Head and Highcliffe there is Christchurch Harbour which is separated from the sea by a low shingle spit. Hurst Spit is another area of low lying ground at the eastern boundary of the Bay, Fig. 3. The cliffs at Hengistbury Head are composed of Bagshot Beds overlain with Plateau Gravel, Fig. 4. The cliffs at Highcliffe are of Barton Clay overlain with Plateau Gravel and Brickearth, Fig. 5, with the Osborne and Headon Beds outcropping toward the east, again overlain with Plateau Gravels. The cliffs are in general soft and very vulnerable to erosion. Cliff recession is especially evident in the central part of the Bay. Erosion of the cliffs is now largely constrained by coastal protection works.

Figure 4. Hengistbury Head, looking west from the Long Groyne.

24. The predominant direction of both winds and waves is from the south west ensuring a nett littoral drift in the bay from west to east. Over many years, the construction of coast protection works has also mainly taken place from the west towards the east and the effectiveness of these works has resulted in the loss of a large proportion of the natural littoral drift. Several key events appear to have been responsible for the present development of the coastline.

(a) In the 1840's a mining company obtained a concession to remove ironstone from Hengistbury Head. Large quantities of ironstone were removed and this appears to have caused severe erosion at Hengistbury Head together with the partial destruction of Christchurch Ledge, a natural offshore feature which may have acted like a groyne and retained a proportion of the littoral drift in Poole Bay.

(b) Subsequent to the mining at Hengistbury Head, there was a rapid growth of existing sand spits at the mouth of

Christchurch Harbour. These spits fluctuated in size and were occasionally breached. The sand forming the spits presumably came around Hengistbury Head from Poole Bay.

(c) In 1938 Bournemouth Corporation constructed a long groyne at Hengistbury Head which effectively reduced the supply of littoral drift material to Christchurch Bay and the spits at Christchurch Harbour subsequently reduced in length. At the same time severe erosion was recorded at Mudeford due to the reduction in length of the sand spit, (Stopher et al, 1966). Successive coast protection schemes where constructed as the process of erosion moved eastward along the coast.

(d) Major works were undertaken by Christchurch Council at Highcliffe during the period 1967 to present, in order to stabilise the coastal cliffs and protect development at the cliff top. Works have also been undertaken by New Forest Council at Barton-on-Sea during the period 1964 to present and at Milford-on-Sea during the period 1930 to 1965. The section of coast between Highcliffe and West Barton, (the responsibility of New Forest Council), is unprotected and now suffering severe erosion due to the construction of works to the west, (the responsibility of Christchurch Council). There is a marked discontinuity in the coastline of approximately 60 metres at the boundary of the two Local Authorities.

(e) The present situation is that, Bournemouth Council has recently completed works to the east of Hengistbury Head as a result of terminal erosion due to the long groyne; Christchurch Council are proposing new works at Mudeford Spit to prevent the spit being breached and New Forest District Council wish to undertake works between Highcliffe and West Barton although the benefit/cost ratio may not be sufficient to obtain an Exchequer Grant.

Figure 5. Barton-on-Sea, looking east. Failed slope stabilisation sheet pile wall shown in the foreground.

COASTAL DEFENCE METHODS

25. The foregoing is a very brief history of protection works undertaken in the Christchurch Bay area and the comments made have the obvious benefit of hindsight. Nonetheless, over many years it appears that with regard to the design of coast protection works the Local Authorities have rated their parochial interests as a priority. This may be due to the manner in which the Coast Protection Act (1949) is interpreted and administered.

MANAGEMENT STRATEGY
26. The main options available to Coastal Authorities with regard to the management of the coastline have been stated by McGown et al, (1988) and can be summarised as follows :

(a) No action (with or without monitoring and most importantly, with planning controls).
(b) Maintenance of existing works.
(c) Renewal of components of existing works.
(d) Construction of new capital works.

27. Each of these has to be considered in relation to the entire Coastal Process Unit in order to assess their consequences.
28. Of particular note is the "No Action" option. In many cases it is the most cost effective and appropriate action for the Coastal Authority to undertake. However, it should involve monitoring the coastal cliffs and implementing planning controls to prevent new developments on potentially unstable sites. This option appears to contravene at least the spirit of the Coast Protection Act (1949), which requires Coast Protection Authorities to prevent coast erosion. However, it can be a very positive and an environmentally sound action.
29. Having regard to the above, a discussion is given below on some elements of current coast protection policy and suggestions are made on how it could be improved in order to have a more effective national coast protection strategy.

Management Units
30. The coastline of the United Kingdom is currently managed by Local Authorities, Water Authorities, Harbour Commissioners and River Boards under the terms outlined by the Coast Protection Act (1949) and the Land Drainage Act (1976). These authorities have permissive powers under the Acts to undertake coast protection and sea defence work. With the exception of Rivers and Harbours, the managed lengths of coast are not related the processes associated with coastal change, but tend to conform to the administrative boundaries of the Coast Protection Authorities.
31. An argument may be made that the coast should be managed in Coastal Process Units, McGown et al (1988), rather than by the Local Authority's territorial limits. These units could be managed either by Coast Protection Boards, provisions for which are available under the Coast Protection Act (1949), or by Standing Committees recognised by Central Government. The main advantage of such a system would be that one authority would be responsible for the management of a Coastal Process Unit and decisions could be made as to the strategy for management of the unit as a whole. Despite the logic of this case, it can readily be appreciated that Local Authorities might well oppose such a proposal on the grounds that they would lose direct control of their parochial interests in coast protection. These parochial interests include environmental and amenity issues related to coast protection works and in particular the impact of these on

tourism. Another area that could lead to problems would be the distribution of the cost of any works between Local Authorities and between Local Authorities and Central Government. Nevertheless the proposal is valid and should be given detailed consideration with a view to resolving these parochial and financial difficulties.

32. Indeed, it is noteworthy that in recent years, on their own initiative, some groups of Coast Protection Authorities, together with other groups interested in the coast, have formed semi-formal committees to discuss problems related to the coast within a specific geographical region. The Standing Conference on Problems Associated with the Coast (S.C.O.P.A.C.) is one group that meets regularly to discuss coastal problems within the Hampshire/Dorset/West Sussex area. The remit of this group is wide ranging and covers "problems associated with the coast" and not just coast protection. However, working groups have been set up to study specific coastal problems. This type of committee has obvious potential, particularly in encouraging liaison between Authorities and perhaps even commissioning and coordinating coastal research within their area. The future for this type of committee can either remain as it is, a liaison group or develop to fulfil some of the previously stated aims of Coastal Process Unit Management.

Planning

33. Local Authorities often appear to have difficulties in preventing development on or near eroding coastal slopes. The problem may, in part, be a result of some of the following factors:

(a) If a developer is refused planning permission to develop a site, an appeal is at times decided in favour of the developer, (by the Planning Inspector or The Secretary of State) without full consideration been given to the long-term stability of the site.
(b) Often the Planning Authority is not aware of the problems associated with the development of potentially unstable coastal sites.
(c) Alternatively, the Planning Authority may not have the expertise to prepare a technically sound case to justify refusal of planning permission.

34. Notwithstanding the above, the use of the existing planning legislation, (Town and Country Planning Act, 1971), with the support of the Planning Authorities, the Planning Inspectors and the Secretary of State, appears to be a suitable and efficient means of regulating capital works for coast protection. Also, the situation with regard to the responsibilities of the Planning Authority and the Developer to development on unstable ground has been clarified by the recent Department of Environment Draft Guidelines (D.O.E. 1988). A key principle suggested in these guidelines is that the Planning Authority should give material consideration to the stability of a site when reaching decisions on a development proposal. The guidelines suggest that, "In the context of the management of coastal zones, Coastal Authorities may wish to consider the introduction of a presumption against built development in areas of coastal landslides or rapid coastal erosion", and, "Development Plans should identify the physical constraint on land within the Plan area". These recommendations, if issued in their present form, should enable Coast Protection Authorities to manage effectively the development of their coastal zone. However, many Authorities will also need technical advice on coastal cliff stability.

35. It is important to realise that the consequences of development adjacent to coastal slopes can also contribute to slope instability.

36. Another factor to be considered is the use of soakaways in rural coastal areas, even when main drainage is available. The main adverse affect of this type of drainage is to cause seepage erosion on nearby cliff faces and to promote soil weathering. An example of this exists, along the North Norfolk coast where soakaways have caused an increase in localised cliff recession. Property owners understandably do not wish to make the financial outlay of connection to main drainage if their property is near to the cliff edge and is of limited value. However, connection to main drainage could improve the life expectancy of the property. This fact is not always recognised by many property owners and so far, few Local Authorities have tried to impose a requirement to connect to the main drainage system.

37. Further, dumping on coastal sites is sometimes encouraged or ignored by Coast Protection Authorities with the view that it is likely to reduce coast erosion. Unfortunately, uncontrolled dumping may also contribute to instability. The deposition of material on the slope may cause undrained loading, (i.e. a short term increase in pore water pressures), or an increase in disturbing forces causing failure. Additionally care must be taken to ensure that deposition of materials on beaches does not adversely affect adjacent lengths of coast.

Capital, Maintenance and Capital Renewal Works

38. Section 21 of the Coast Protection Act (1949) allows the Secretary of State to allocate Exchequer Grants to approved coast protection works, always subject to such conditions as the Treasury determine. This section does not distinguish between capital and maintenance works, although it has been the policy of the Ministry of Agriculture, Food and Fisheries, (and previously of the Department of the Environment), only to grant aid approved capital works. Maintenance works have been funded by the Coast Protection Authority. Slope stabilisation work is often included as part of the coast protection scheme when a new sea wall is built or where the stability of a existing wall is threatened by cliff slippage. However, slope stabilisation work in-isolation has not attracted grant aid.

39. A coast protection scheme may include many component elements, such as; beach, groynes, sea walls, slope drainage, slope regrading, planting of vegetation, hinterland drainage, etc. These components interact to form the coast protection scheme. Some of the elements may be regarded as one off items of capital expenditure, e.g. the slope regrading. Some of the other items may have a finite life less than the design life of the scheme and need to be renewed in order to maintain the design performance. For instance, a sea wall may have a design life of 120 years, whereas slope drainage may only have an effective life of 20 years. Although the complete scheme may initially attract grant aid, the renewal of the individual components is regarded as maintenance unless the stability of the "coast protection" is threatened. The inequality of effective life of components of the coast protection scheme puts the burden of component capital renewal on the Coast Protection Authority. Inevitably many Authorities only do "break-down" maintenance, which can lead to accelerated deterioration in the performance of the complete works. It is suggested that consideration be given by Central Government to the introduction of Capital Renewal Grants to enable the Coast Protection Authorities to replace components of the coast protection scheme.

Detailed investigation for coast protection works
40. Recently, the technical requirements placed on Coast Protection Authorities in order to justify coast protection works have become more stringent and the financial resources available for coast protection have been reduced over the past few years. In order to justify works the Coast Protection Authorities need to present a sound technical case. Often this involves a long-term in-depth study of the causes of coast erosion and the consequences of any proposed works. The outcome from such a study may recommend; do nothing, carry-out maintenance, or construct new works. However, it is only for the latter case that the cost of the study is likely to attract an Exchequer Grant, and then only if the works are approved. A thorough long term investigation for proposed coast protection works can be costly and there appears to be no incentive for the Authorities to consider the alternative options to new capital works. Consideration should therefore be given to the feasibility of providing grant aid for detailed investigations of coast protection works. This might be considered once the Authority has carried out sufficient preliminary work to justify the need for a more detailed study. Provision for this type of investigation is available under the Land Drainage Act (1976), (MAFF, 1976), but not under the Coast Protection Act (1949). With a change in the appropriate legislation, this approach may prove to be highly cost effective in relation to overall expenditure on coast protection works.

Monitoring
41. The performance monitoring of existing coast protection works and of unprotected lengths of coast varies considerably form one Coast Protection Authority to another. Some Authorities carry out virtually no monitoring whilst others monitor only components in the system, e.g. beach levels, slope movements or pore water pressure.
42. The integrated monitoring of the overall coastal process system is a vital component of coastal management. Its benefits can enable the Coast Protection Authority to programme maintenance work, highlight problem areas, judge the performance of existing works and improve the design of future works. The coordination of coastal monitoring at a local, regional and national level should be an important aspect of any Coastal Management Scheme. For example, monitoring work is often undertaken, say by a Water Authority on beach levels, but the data is not in a usable format for a Local Authority. This may lead to the duplication of monitoring or the latest data not being used for the design of works. The investment made in coast protection works warrants the expenditure on monitoring. The work undertaken to date by the Coast Protection Authorities has not received direct grant aid. To encourage further monitoring by all Coast Protection Authorities, it could be insisted that monitoring work be undertaken as a condition of approval for grant aid. Also there appears to be the need for national coordination of the storage and interchange of coastal monitoring data.

CONCLUSIONS
43. The main problems associated with coastal landslides in the U.K. are within the Quaternary and Tertiary deposits that form much of the south and east coasts of England. The principal cause of coastal landsliding is the erosion by the sea, but failures do also occur on protected slopes.
44. There is a need for expert geotechnical advice on the design of coastal slope stabilisation works.

45. Consideration should be given to changing the present system of managing coastal areas away from individual Coast Protection Authorities to either Coast Protection Boards or Standing Committees recognised by Central Government.
46. Initiatives such as S.C.O.P.A.C. should be encouraged, formally recognised and recommended to other groups of Coast Protection Authorities.
47. Planning controls should be used to manage the development of coastal zones, once potential hazards have been identified.
48. Consideration should be given to the provision of Exchequer Grants for capital renewal and detailed investigation works.
49. Coast Protection Authorities should be required to undertake maintenance and monitoring work as conditions for grant aid.
50. National coordination is required of monitoring data on coastal processes.

ACKNOWLEDGEMENTS

The authors wish to acknowledge the support of the Ministry of Agriculture, Fisheries and Food who have provided research funding for the work reported in this paper. The views expressed in the paper are those of the Authors and not necessarily those of the Ministry of Agriculture, Fisheries and Food.

REFERENCES

ANON. Local Echoes, The Scarborough Gazette, May 28th, 1896.
CHANDLER R.J. Delayed failure and observed strength of first-time slides in stiff clays: a review. Proc. IVth Int. Symp. Landslides, 2, 19-25. Toronto, 1984.
D.O.E. Minerals planning guidance: Development on unstable land (draft). Department of the Environment. 18, Aug. 1988.
HERLIHY A.J. Coast protection survey, 1980. Department of the Environment, 138, London, 1982.
HUTCHINSON J.N. and GOSTELOW, T.P. The development of an abandoned cliff in London Clay at Hadleigh, Essex. Phil. Trans. Roy. Soc. London, A 283, 557-604, 1976.
M.A.F.F. Review of land drainage grant arrangements. Introduction of revised memorandum relating to grants. Memorandum by the Ministry of Agriculture, Fisheries and Food, LDW 29088A, Aug. 1976.
MCGOWN A., ROBERTS, A.G. and WOODROW, L.K.R. Long-term pore pressure variations within coastal cliffs in North Kent, U.K. 9th Euro. Conf. S.M.F.E. 1, 455-460, Dublin, 1987.
MCGOWN A., ROBERTS, A.G. and WOODROW, L.K.R. Geotechnical and planning aspects of coastal landslides in the United Kingdom. Vth Int. Symp. Landslides, 2, 1201-1206. Lausanne, 1988.
SKEMPTON A.W. and BROWN, D.J. A landslide in boulder clay at Selset, Yorkshire. Geotechnique, 11, 280-293, 1961.
STOPHER H.E. and WISE E.B. Coast erosion problems in Christchurch Bay. Journal of the Institution of Municipal Engineers, 328-332, Oct. 1966.

18. Coastal structures

N. PALLETT, FICE, MIHT, MConsE, FBIM, and S. W. YOUNG, BSc, MICE, MIWEM, Dobbie and Partners

SYNOPSIS Coastal structures must be carefully conceived and designed in order to ensure that there can be an appropriate level of confidence that they will achieve their design purpose. The choice of structural form and constructional materials must take full account of the forces of the sea to which the structure will be subjected and the regime within which it is to be located. This paper identifies those aspects of the coastal environment that will particularly influence the design of coastal structures and outlines the constructional constraints that must be given proper consideration in the development of solutions to coastal problems.

INTRODUCTION
1. For the purposes of this paper, coastal structures are defined as artificial constructions designed to form part of a coastal management strategy such that an adequate defence against incursion by the sea is provided or coastal erosion is limited.
2. Coastal structures may take many forms but in this paper are considered under three separate categories:-
(a) Sea walls, where the structure is designed to form the landward boundary of the sea's activity such that erosion or flooding of land and property behind the wall is restricted.
(b) Beach management structures, which may be constructed in association with or in place of a sea wall, but whose primary function is to maintain the beach as part of an overall sea defence or coast protection strategy.
(c) Other coastal structures, such as interceptor drains, which may be constructed in association with sea walls or beach management structures, whose purpose is to limit the effects of the sea once it has passed the primary coastal defence or at the shoreline.
3. Whilst the various types of coastal structure are required to fulfil different roles, the factors that will influence their design and construction are similar. Any structure must be designed with due regard to the overall coastal regime and should be perceived as part of a complete coastal management strategy.

COASTAL DEFENCE METHODS

4. Design methods and construction techniques have advanced considerably in recent years, allowing the coastal engineer to develop alternatives to the solutions that have traditionally been adopted for coastal problems. It is important to note, however, that whilst these improved design methods allow a reasonable level of confidence to be developed of the way in which a coastal structure will perform, many of the design practices still rely on empirical methods and engineering judgement.

THE COASTAL ENVIRONMENT

5. The environment in which coastal structures are located is one that is in a perpetual state of change. In order that the coastal structure will be able to perform satisfactorily it is necessary to assess the causes and likely magnitude of these changes both in the short and long term. This assessment will normally take the form of a detailed study of the coastal hydrodynamics of the frontage, preferably nested into a strategic study for the region of interest.

6. Any structure introduced into the coastal zone will have an effect on the overall regime of the foreshore. Whilst this effect may be the design purpose of the scheme, such as the stabilisation of the beach or the prevention of the landward movement of the shoreline, there are also likely to be other effects which may be adverse and to which due consideration must be given in the design process. These adverse effects may include increased seaward reflection of wave energy, causing the loss of beach material in front of the structure, or a reduction in the supply of material moving along the coastline to adjacent frontages. In the longer term the construction of a coastal structure may introduce discontinuities into the foreshore alignment as a result of the movement of adjacent stretches of coastline where structures have not been constructed. In some instances the supplementary effects of the construction of a coastal structure may be beneficial but it is also necessary to have evaluated these effects during the design of the structure in order that full account can be taken of them when assessing the feasibility of the proposals.

7. The above complex interactions with the existing coastal regime will differ for various forms and alignments of the coastal structure and a wide diversity of choice exists at the conception stage of a solution to a coastal problem. In order to assess the optimum form of construction, the design must be undertaken interactively with an assessment of the effects of the alternative construction forms on the nearshore processes. In many instances, the adoption of beach recharge techniques may be an alternative to the construction of coastal structures or may be considered as a supplement to them. This will further increase the number of alternative solutions to be considered.

8. Tidal and wave considerations are fundamental to the design of coastal structures both because of their hydraulic effects on the structure and because they will significantly affect the materials that can be specified and the construction methods that can be adopted. Records of waves and tidal levels for coastal sites are frequently only of short duration and must be extrapolated to predict extreme conditions, allowing for secular rise in sea levels and atmospheric changes. Research is currently continuing into the joint probabilities of waves and water levels but no standard methods have yet evolved. Global increases in sea levels must now be anticipated during the design life of most structures being conceived at the present time. Whilst predictions are not sufficiently accurate for structures to be designed to precisely accommodate the effects of this change, it must be given due consideration in the design process and structure concepts should be sufficiently adaptable to cater for the anticipated changes in sea level.

9. There are few occasions when coastal structures can be designed purely on the basis of hydraulic and regime considerations. The coastal environment is highly valued as an important resource, providing unique natural habitats and facilities for recreation. In the development of coastal structures, these aspects of the coastal zone must be carefully considered and appropriately taken into account if the structures are properly to serve the community as a whole.

10. The coastal environment is particularly harsh and aggressive and coastal structures must be designed to withstand the forces to which they will be subjected. This will restrict the choice of materials available to the designer. Abrasion and corrosion will be key considerations together with tidal and wave influences on construction.

11. It is important too in the design of coastal structures to recognise the empirical nature of the design methods that are used and the incomplete nature of our understanding of the marine environment. Traditionally, a conservative approach to the design of coastal structures has been adopted and inadequacies of understanding of the coastal regime have tended to be compensated for by this approach. This traditional approach has not always been successful. Failure to take into account the dynamics of the coastal regime either due to ignorance, inability to analyse or a cavalier attitude by adjacent frontagers has led to many sea walls not being able to perform satisfactorily for the whole of their design life. With improvements in our ability to evaluate the dynamics of the nearshore processes, we must be careful to continue to pay due regard to the inadequacies in our understanding and to allow for this in our design philosophy.

COASTAL DEFENCE METHODS

DESIGN METHODS

12. For many years the top level of sea walls was set on the basis of Mean High Water Springs plus between 9 and 12 feet dependant on the site's exposure, beach and other conditions and the risks associated with overtopping. This type of approach worked reasonably satisfactorily together with the design of groynes based directly on previous experience. This experience is insufficient on its own to evaluate the wide variety of solutions that are now available for consideration for any scheme and to allow an appropriate level of confidence to be attached to the design, particularly where the scheme location is unusual.

13. In many ways, the types of coastal structures that are now able to be considered are a direct product of the advances in the design methods that have been made over recent years. Computational speed and efficiency now mean that detailed wave records can be analysed over long time series. Large scale representations of the sea bed can also be handled mathematically to allow detailed wave refraction analyses to be undertaken. From the inshore directional wave spectrum thus derived, beach models can be driven and the interaction of coastal structures with the coastal hydrodynamics can be predicted, although often in an idealised way.

14. Mathematical modelling techniques are continually advancing, but at the present time they have difficulty coping with rapidly changing scenarios such as that of a shingle beach between groyne bays with varying incident wave directions. Whilst computational speed has increased there is still considerable effort involved in the analysis of beach movements using "multi-line" techniques. The more frequently used method is therefore a "one-line" model where the beach is assumed to move in parallel increments at each step, each position of the whole beach therefore being able to be described as a single line. This type of technique is limited in its application but for many situations provides a sufficiently accurate representation of a structure's interaction with the beach such that its overall performance can be assessed. Mathematical modelling of this type may then be supplemented by a more sophisticated mathematical model or a physical model.

15. For the hydraulic design of coastal structures it is important to consider wave interaction with the structure in terms of overtopping, energy dissipation and reflection. Each of these characteristics may be evaluated in approximate terms by empirical design methods, although considerable experience is necessary in the application of these methods as frequently they will require to be interpreted or extended to suit particular design circumstances. In many cases the empirical formulae will have been derived for regular waves in the laboratory and will need to be reviewed in order to obtain estimates applicable to real sea states. For many

structures a physical model will be the best method of confirming the results of the empirical hydraulic design.

16. Physical modelling is now more flexible than in the past. Measurement accuracy and general techniques have improved such that reliable results are able to be achieved more readily and at realistic cost. The construction of a physical model to verify the mathematical model results or a basic design calculation is now a sensible consideration for any sizeable scheme with any degree of complexity. The model to be used may be a simple two dimensional flume model in order to check a sea wall profile. Alternatively, or in addition, a full basin model may be used to check a beach management strategy, the behaviour of an offshore structure or sediment transport patterns in the vicinity of complicated plan shape discontinuities. Physical models do have limitations in terms of the number of sea states that can be analysed and the complicated relationship between sediment transport in the model and that which could be expected on the foreshore itself. Nevertheless, in many situations a physical model will be an appropriate design tool in the assessment of the behaviour of a coastal structure.

CONSTRUCTION

17. The coastal environment has been described previously and some of the difficulties of working within the coastal zone have been indicated. Tidal movements will restrict working hours and the forces that will be expended on partially completed works or freshly placed materials may be severe. In addition, the locations and nature of coastal structures often introduce difficulties during the construction stage. Coastal construction sites will generally be linear in form and the area free from tidal action is likely to be small. Operations will therefore be restricted and it is often not possible to work on a number of constructional sections at one time without introducing undesirable congestion on the site.

18. Access to coastal sites is also often very difficult. Many estuarial walls are located on poor ground surrounded by low lying areas over which access for all but the lightest of plant may be impossible. Even where coastal structures are located close to centres of population the limited vehicular approach directions mean access routes to the site frequently pass through the centres of the towns. The importation of materials to the site is therefore often restricted and the access routes are frequently congested during the summer months by tourists and holiday traffic. In many cases construction operations on the foreshore are even suspended during the summer months which is usually the most active period in the construction industry.

19. Sites located at the base of cliffs will be even more restricted in nature. In some cases constructional costs will have to include the cutting of an access route from the

top of the cliff. In other instances, the use of marine plant may be considered. Provided there is sufficient tidal range and the foreshore is sufficiently smooth the use of floating plant to import material and undertake construction operations can often be financially attractive. This is particularly so in the case of large rock from sources such as Scandinavia, where the fiord side location of many quarries means that the rock barges or ships can be very easily loaded with material.

20. For situations where construction operations are required to extend to below the low water mark, construction by marine plant may be the only sensible option. Construction in these circumstances and locations does, however, require considerable knowledge by the designer of the available floating plant and its capabilities in order that the designed coastal structure is readily able to be built. Improved real time forecasting of waves and tidal levels is now making the use of floating plant on the UK coastline less hazardous, although the distance to an appropriate sheltered haven can significantly affect the available working hours on the site.

MATERIALS

21. A recent survey of existing sea walls indicated that the most predominant existing constructional material is concrete. Concrete has been widely available for a large number of years and has shown itself to be a very versatile material that is able to fulfill many of the requirements demanded of a coastal structure.

22. For concrete to be durable in the marine environment it must be properly specified and detailed and an appropriate method of working must be adopted. The major problem that has been identified with concrete structures in coastal works is abrasion by shingle under wave action. To resist this abrasion the aggregate must be specified to be as hard as possible and at least of equal hardness to the beach material. It is also necessary for the designer to ensure that an adequate cover of dense concrete is achieved to any steel embedded in the concrete to prevent corrosion.

23. It is recognised that the curing stage of the concrete is important to its long term durability. The requirement to achieve good control at the curing stage has strengthened the case for the use of precast concrete in the marine environment as precasting provides the ability to isolate the wet concrete from the marine environment.

24. Precast concrete allows the Contractor to undertake the fabrication of pieces of the finished work in a carefully controlled manner remote from the influences of the tide and the sea water environment. An extremely high standard of surface finish is also able to be achieved using pre-casting, thereby minimising any requirement for remedial works and allowing badly formed pieces to be readily rejected. In many

cases, the application of a nominal pre-stress to concrete sections has been found to be beneficial, restricting any tendency for cracks to form in the concrete work and hence increasing the durability of the finished structure.

25. Good design of concrete for the marine environment will ensure that sharp corners and arisses are avoided since these are particularly vulnerable areas of the construction. Joints and their sealants are also a further area that are particularly susceptible to damage and careful detailing of these components is necessary in order to ensure that they do not constitute a continuing maintenance problem.

26. Iron and steel are sometimes used as major components of sea wall structures and in some areas may form the sea wall itself. Common applications are as foundation piles, in steel sheet piling or as connecting pieces between elements of the structure. The salinity of the marine environment will cause corrosion of iron based metals to be rapid in the intertidal and splash zones. Where abrasion occurs the life of steel plate is likely to be particularly short and in general terms the use of steel on beach structures subject to abrasion by wave born sediments should be avoided.

27. In some locations around the country steel sheet piling was used for the construction of groynes. Where they remain buried in the beach the piles perform satisfactorily, but as soon as they become exposed to abrasion, deterioration takes place very quickly. The jagged form that these piles then assume is a very severe hazard to beach users and the use of sheet piles in groynes, particularly on shingle beaches, is not desirable. Hardened steel piles have been used in this application but their service life is not significantly extended and they are now no longer manufactured.

28. Timber was frequently used in historic times to form the support piles to sea walls, although its use in the exposed sections of walls has been less common. Breastwork type protection has been constructed using timber, particularly when an open form of construction is required that will allow some material to supply the alongshore transport requirement of the coastline. It is in groynes, however, that the use of timber has been widespread both from local and foreign sources, although imported hardwood is probably now the most widely used timber for this application.

29. In selecting timber for coastal structures it is important to give consideration to its susceptibility to attack by marine borers. Fungal decay of the timbers can be a problem but it has generally been found that the saline marine environment has inhibited the fungal decay of exposed timbers whilst driven timbers will be effectively isolated from oxygen by the ground, so decay will also be limited.

30. If an appropriately durable species is selected, the design life of timber in the coastal environment can be long.

COASTAL DEFENCE METHODS

Timbers such as the imported hardwood Greenheart are very resistant to abrasion and other forms of degradation and have performed very satisfactorily in use for many years. Timber is particularly suitable for groyne construction, where it may be necessary to amend the planking levels and profile during the life of the structure, since such adjustments can be made without difficulty.

31. Large rock is currently being widely used for the construction of coastal structures in this country. As a naturally occurring material it is often favoured in environmentally sensitive areas. The large voids and uneven surface of the material means that it is not well suited to an amenity beach, but in less intensively used locations it can find wide public acceptance. The open nature of an armourstone or riprap apron has good energy dissipating properties. Much research has been carried out on the sizing and placing of large rock to resist wave forces for breakwater construction and this research can be used to provide a basis for the design of nearshore coastal structures.

32. The ability of large rock to be placed below the low water mark has meant that it has been used in the construction of offshore breakwaters, strong points and large groynes. This characteristic together with its energy dissipating and other properties mean that it is likely to be used even more extensively in the future. In particular on a number of schemes where falling beach levels have exposed the toe of an existing structure, large rock is being used to extend the wall toe.

33. Asphaltic products have been used in the construction of sea walls for many centuries primarily as a jointing material between the stones or concrete blocks of revetments. Asphaltic concrete can also form a strong, abrasion resistant material for the construction of sea walls and in recent years the permeable nature of underfilled asphaltic products has started to be exploited. Such materials are lean sand asphalt and open stone asphalt which have been used both in sea wall and groyne construction.

34. The tidal movement of water and the elevation of adjacent land masses often causes significant groundwater movements beneath coastal structures. The use of a permeable material will mean that the sea wall does not have to be designed to resist the pressures created by these movements. Open stone asphalt is appropriate in this application and in addition, when used to form the upper layer of a sloping structure, will dissipate small wave energies within the core of the material, thereby reducing wave run-up and backwash velocities and encouraging the deposition of material at the toe of the revetment.

35. Gabions, geotextiles and other light materials may sometimes find application in the construction of coastal structures. Their application, however, will generally be

limited to those sections of the structure not directly exposed to severe wave action.

SEA WALLS

36. Since ancient times walls have been constructed to resist the sea's attack or to prevent the inundation of coastal lands. The constructional forms were initially limited by the available materials from which a choice was made according to the degree of exposure of the site. Earth banks were first used to resist the sea's inundation in estuarial type situations but in all but the most sheltered of locations these required protection from wave attack. This protection was first provided by vegetation strengthened by timber and brushwood. Stone pitching and masonry protection were also used over some lengths either to provide increased protection or for aesthetic reasons. Protected earthen banks of this form were generally recognised not to be appropriate on coastlines where significant wave attack was expected. More robust techniques were introduced in these locations. Masonry was the preferred construction material in many such situations. Traditional building techniques were adopted and the walls were generally near vertical with rubble hearting.

37. The vertical form of construction was compatible with many of the other constraints on wall design. Preferred construction methods meant that it was desirable to carry out the works as high up the beach as possible in order to limit tidal effects and provide maximum protection to the working area. Additional land acquisition for a sloping wall behind the beach was not encouraged and wall construction tended to be confined to a narrow strip of land. The theory for the hydraulic design of sea walls was also not well developed, and it was therefore not possible to put forward a strong case for the use of flatter front slopes for sea walls until relatively recent times. These constraints and limits of understanding of the hydraulic performance of sea walls and beach changes explain why, until the middle of this century, the most common form of construction for sea defences was of near vertical walls located at the top of the beach.

38. Whilst it was recognised that the purpose of the sea wall was to prevent wave energies from affecting the area landward of the wall, it was not always recognised that unless the wave energy was dissipated it would merely be reflected seawards, often with severe consequences. Reflected waves will increase the energy levels in front of the wall causing additional material to be put into suspension and sometimes increasing the alongshore transport forces. The design of sea walls now places considerable importance on limiting wave reflections wherever possible and encouraging the dissipation of wave energies in a manner that is not detrimental to the foreshore.

39. The more reflective forms of sea walls such as simple

COASTAL DEFENCE METHODS

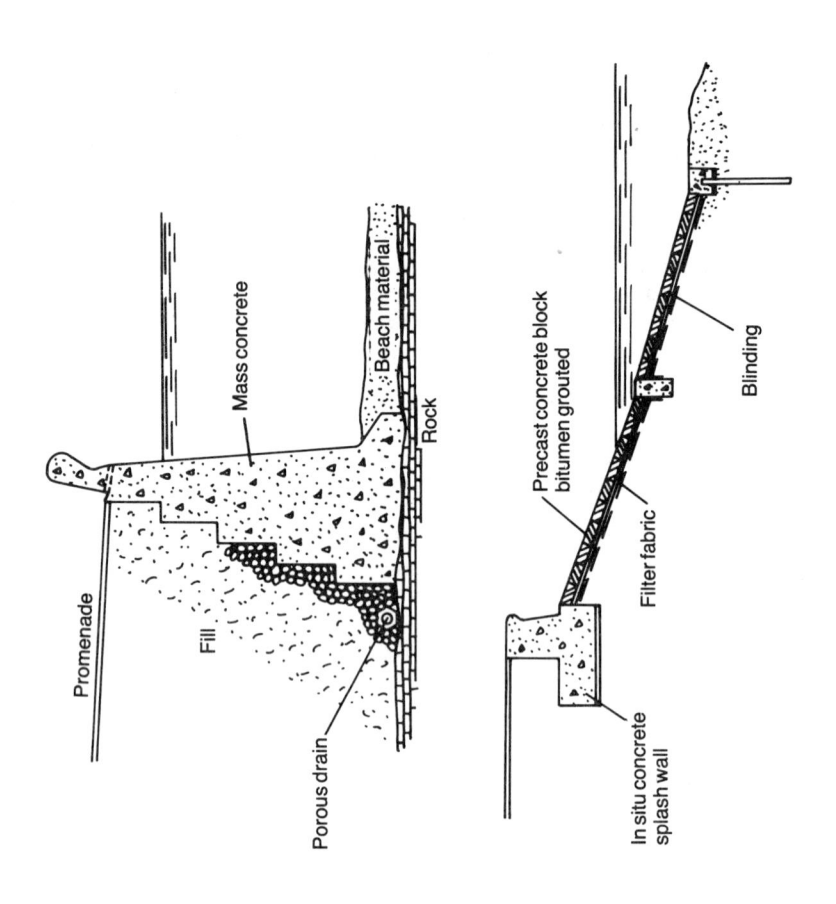

PAPER 18: PALLETT AND YOUNG

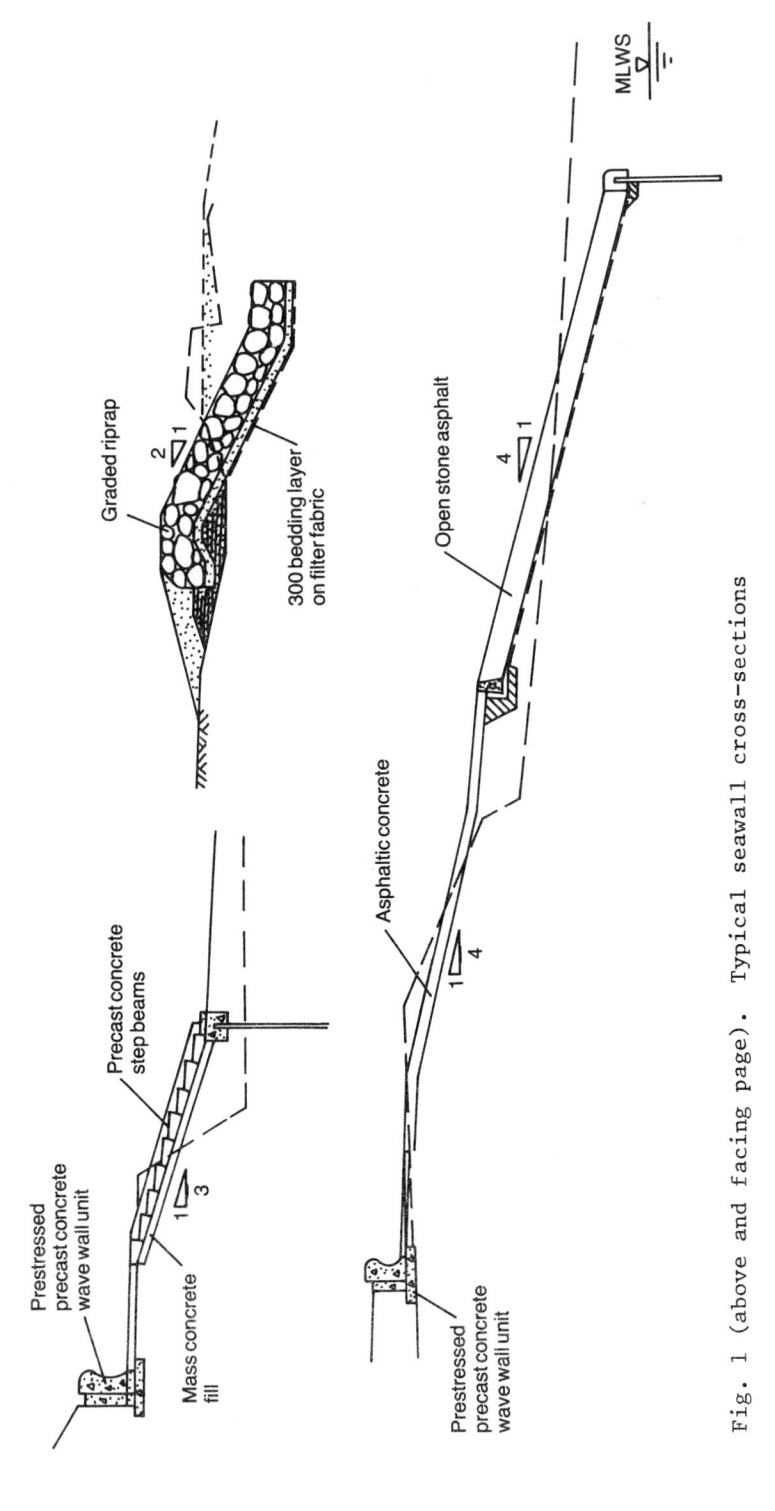

Fig. 1 (above and facing page). Typical seawall cross-sections

vertical walls have therefore generally given way to sloping constructions in those areas where it is important to retain a beach at the foot of the wall. Experience has shown that, whilst the flatter a slope is the less energy will be reflected, a reasonable compromise slope that will satisfy the other constraints in most locations is between 1 in 4 and 1 in 3. Where other considerations permit, further energy dissipation may be achieved by adopting porous upper surfaces to the sloping wall face. With this type of construction the wave energies are reduced as the water is forced to flow both over and through the body of the structure and its underlayers.

40. Graded riprap, armourstone and randomly placed concrete armour units have often been adopted to achieve this dissipation of energy. The voids which allow these materials to dissipate wave energy within their structure may also present a considerable hazard to beach users in areas of high amenity usage. In such locations the more hydraulically desirable forms of construction may therefore have to be avoided in preference to those which are more compatible with the general public's use of the area. Breakwater research has led to the use of pattern placed open concrete units in place of armourstone in some locations. Relatively light units have been shown to be capable of replacing large armourstones and their energy dissipating properties are very attractive. In some coastal structures blocks of this type have been successfully used but their voided upper surface still presents a considerable hazard on an amenity beach.

41. In spite of the merits of porous construction, in many locations it will therefore be necessary to adopt a revetment surface that is not porous. Concrete stepwork often provides a satisfactory constructional form in areas of high amenity usage since the steps allow access to be taken along the whole length of the wall. Although adding some roughness to the upper surface of the slope, the steps cannot be taken as making the surface rough in hydraulic terms. Under large wave attack the step risers will fill with water effectively forming a sloping revetment. Some experiments have been carried out introducing slots through a stepped seawall to a porous underlayer. Whilst this type of structure may reduce small wave energies and encourage material accumulation on a sandy beach, under large wave attack it cannot be considered as being sufficiently porous to dissipate large wave energies.

42. Open stone asphalt as described in the materials section of this paper has similarly proved effective in dissipating small wave energies and encouraging sand accumulation.

43. The reduction in reflections of wave energy from a sloping wall has been outlined. It may not be possible or even desirable to continue this type of sloping construction to a height sufficient to limit overtopping of the structure.

Many sea walls therefore incorporate a wave wall at their crest. Located well above the beach the function of the wave wall is to contain the wave energies originating from waves breaking on the body of the wall. Wave walls are therefore frequently constructed with a recurved front face to return this water seawards causing it to lose energy as spray. At high still water levels the wave wall may reflect some small waves, but with sufficient depth of water, these reflections should not be too detrimental to the frontage. The recurved shape commonly adopted is amenable to concrete construction which can usually be undertaken at this level without undue difficulty and is capable of receiving surface treatments to match its surroundings.

44. Research has shown the benefit of introducing a berm between the sloping body of the wall and the wave wall since this allows the height of the wave wall to be reduced without an increase in the overtopping discharge. This form of wall is therefore frequently adopted, since the berm may also form an important amenity benefit as a public access route.

45. It is important in the design of a sea wall to ensure that the structure is provided with an adequate toe. Recent surveys undertaken by the Construction Industry Research and Information Association have shown that the most common cause of sea wall failure originates at the wall toe. In order to overcome this problem, it is essential that the toe of the structure is extended to an appropriate depth. It is also important that the structure is conceived from a knowledge of the coastal hydrodynamics of the foreshore and that, if necessary, the wall is accompanied by beach control or other works to ensure an adequate beach is retained. An appropriate level of beach monitoring is then also required by the Authority responsible for the structure in order to ensure that appropriate measures are initiated in good time if the beach behaviour is other than that predicted for the design of the structure.

46. The Construction Industry Research and Information Association is currently funding the preparation of Sea Wall Design Guidelines. These comprehensive guidelines are intended to give guidance on the design of sea walls with regard to the philosophy and techniques of design, the factors that should be considered and how they should be evaluated, as well as describing the statutory obligations and procedures.

BEACH CONTROL STRUCTURES

47. Beach control structures are defined for the purposes of this paper as those coastal structures whose primary purpose is to retain beach material on a frontage. Some structures such as offshore breakwaters may fulfil the additional role of dissipating or reflecting wave energies such that the requirement for a sea wall may be reduced but these structures will be included in this section. Other

COASTAL DEFENCE METHODS

beach control works are primarily of the groyne type, constructed approximately at right angles to the shoreline, although some such as artificial headlands are designed to interact with the incoming waves in rather a different manner to a normal groyne field.

48. The standard type of groyne that has been widely used in this country is of the fence type, consisting of a vertical faced structure probably of hardwood but sometimes of concrete or steel sheet piling extending across the beach. The groynes are designed to control the movement of beach material along the foreshore. Various lengths of groyne have been adopted. Short groynes are recognised as being applicable on shingle beaches whilst longer groynes have been used on sandy beaches, but generally with less success. The vertical faces of 'fence' type groynes will cause reflections of wave energy within the groyne bay that may not be beneficial to the accumulation of fine material.

49. Alternative materials for the construction of groyne structures have therefore been adopted in some instances. Groynes of asphaltic construction have been used in the Netherlands and on the Lincolnshire coastline. Generally in this country, however, where large fluctuations in beach level are experienced alternative constructional forms have tended to be of rock. As with seawalls the ability of rock to dissipate wave energies within the structure of the material is recognised as reducing energy levels within the groyne bay and encouraging the accumulation of beach material.

50. Observations have also shown that within groyne bays it is often possible for seaward currents to occur along the faces of the groynes. The uneven faces of large rock groynes tend to reduce the velocities of these currents, further aiding material retention within the groyne bay. The flexible nature of rock construction also allows the profile of a rock groyne to be amended if necessary and is to some extent self-healing if beach movements are greater than anticipated. To provide additional sheltering within groyne bays and encourage deposition adjacent to the groyne sides rock groynes may be constructed with T or Y shaped heads. The head is designed to encourage the dissipation of wave energy on its front face and as such the use of a porous medium such as large rock has much to commend it.

51. The concept of an artificial headland attempts to overcome the fact that groyne structures do not normally accumulate material on their leeside. The artificial headland is usually formed of large rock pieces with a head that is designed to encourage wave diffraction and a wave height gradient into its lee as occurs with a natural headland. These structures however have only found limited use since, although extended to below the low water mark, they are not on the same scale as natural headlands and have little diffraction effect on longer waves. Constructions of

this type extending to well below low water will also tend to cause a more significant interruption to the alongshore movement of material than simple groynes. Whilst they can be designed to allow some material to pass over their stems, this is generally difficult to achieve and they may divide the coastline into discrete cells which will not be naturally recharged.

52. Various combinations of these design principles have been adopted in the design of other beach control structures. These structures however, are generally at the prototype stage and in most cases have not been proved under a wide variety of wave and tidal conditions over a long time period. The coastline is divided at some points by natural barriers to alongshore transport such as strong estuary mouths and in some instances a natural beach terminates quite abruptly at a change in coastal physiography. At these points terminal structures may be constructed to contain as much of the natural movement of material as possible. The material thus accumulated is then available for recycling or recharge to an adjacent coastal cell.

53. Offshore breakwaters have been widely used on the coastlines of Spain, France and Italy to retain healthy beaches, although often in locations where the incident wave conditions are limited in period and direction. They have also been used in Japan and in a limited number of places in this country. However, the large tidal range and strong alongshore movement of material around much of our coastline means that it is not possible simply to transpose the experience of other countries to our shoreline.

54. Breakwaters may be either submerged or surface piercing. In the design of breakwaters that are always submerged some economy may be made in the constructional materials that are used but their effects on the incident wave climate will obviously be limited at high tidal states. Sheltering by an offshore breakwater will cause a reduction in alongshore transport and wave energies in its lee and appropriate locations for the construction of such structures will be where they are required to provide shelter to a particular short stretch of coastline or to spread erosion at the boundary between a sea wall and an unprotected coastline.

55. A choice must also be made on the provision of a shore connection to the breakwater. Tidal currents may be concentrated by the construction of an offshore breakwater causing an inshore scouring action that may negate the beneficial beach building characteristics of the structure. In these instances, a shore connection may be adopted. The level of this shore connection will be required to be carefully set in order that it achieves its design purpose without completely obstructing the alongshore movement of material where this would be undesirable for the frontage. In view of their energy dissipating role offshore breakwaters have usually been constructed of large stone or randomly

placed concrete armour units, although pattern placed open concrete units have sometimes been adopted.

OTHER COASTAL STRUCTURES

56. Seawalls and beach control structures are by far the most predominant types of coastal structures. It should be noted however that alternative solutions have been adopted at some locations. Such a solution may take the form of a large interceptor drain. Instead of the traditional seawall structure which limits overtopping this alternative philosophy allows water to pass over or seep through the primary defence to then be routed away before it has opportunity to cause flooding. Solutions of this type can only be adopted where there is either available storage for the flood water or a low level discharge to calmer water can be achieved. In these instances a structure of this type can be made very unobtrusive and may well fulfill the other design objectives of the scheme.

57. Coastal structures designed to reduce wave activity at the shoreline for the purposes of reducing the requirement for a sea wall have been conceived but not widely adopted. Structures of this type are basically similar in performance to the offshore breakwater but have been particularly conceived for areas where low cost works are required. They have therefore been designed to be constructed from inexpensive materials such as old rubber tyres or even redundant car bodyshells. Seaweed and other floating materials have been conceived with similar aims but have not been applied practically on any scale.

58. An alternative to the construction of a seawall to prevent coastal cliff erosion has been the introduction of a cliff drainage system. The opportunities for this type of system, however, are limited according to the location of the site.

ACKNOWLEDGEMENTS

Our grateful thanks are expressed to Dr A H Brampton of Hydraulics Research Ltd for his helpful comments on the draft of this paper.

REFERENCES

1. Sea Walls - Survey of performance and design practice, CIRIA Technical Note 125, 1986.
2. Sea Wall Design Guidelines, compiled by Posford Duvivier, CIRIA/Butterworths 1989 (in press)
3. Brampton A H and Smallman J V Shore protection by offshore breakwaters, Hydraulics Research Ltd Report No. SR8, July 1985.
4. Berkeley Thorn R and Roberts A G Sea Defence and Coast Protection Works, Thomas Telford Ltd, 1981.
5. Summers L and Fleming C A Groynes in coastal engineering: a review, CIRIA Technical Note 111, 1983.

19. Scheme worthwhileness

E. C. PENNING-ROWSELL, A. COKER, A. N'JAI,
D. J. PARKER, and S. M. TUNSTALL, Flood Hazard Research
Centre, Middlesex Polytechnic

SYNOPSIS. The returns derived by society from investing in coast protection and management should determine the scale of any works anticipated and constructed. Both government and coast protection agencies are becoming increasingly concerned about the efficiency of investing scarce resources on coastal management and therefore about scheme worthwhileness. This paper discusses a number of examples where the benefits of coast protection schemes have been researched, and presents insights into the problems that such benefit-cost analyses contain. Given that coast protection works are often expensive, the research shows that scheme design must be tailored to likely benefits, rather than benefit-cost analysis being used retrospectively to justify pre-determined designs.

INTRODUCTION

1. In 1985 the responsibility for the oversight and grant-aiding of coast protection schemes in England was transferred from the Department of the Environment to the Ministry of Agriculture, Fisheries and Food. The Ministry has traditionally been responsible for oversight of sea defence against flooding, so this move should allow an integrated approach to investment decisions concerning coastal management. To help this process of investment appraisal the Ministry has comissioned the Coast Protection Research Project (COPRES) at the Middlesex Polytechnic Flood Hazard Research Centre. This research is designed to improve the data and methods for gauging investment worthwhilenes in this field.

2. This COPRES research project is investigating the value of a number of coastal resources protected from both erosion and flooding, and hence the worthwhileness of the necessary levels of investment. This worthwhileness is a function of the property to be damaged or lost, and the recreation, amenity and environmental losses anticipated with a continuation of coastal erosion and flooding (ref. 1). Once these losses have been determined and calculated as a capital sum worthwhile investing, the engineer and others should attempt to fit any proposed scheme into this budget total so as to minimise residual losses.

3. The process of benefit assessment, and hence budget determination, is complicated by the methodological and technical problems of assessing amenity and environmental gains from coast protection and management, including the number of 'intangible' elements involved. The calculation of property losses is relatively straightforward, albeit complicated by the need to determine erosion lines and agree future property values, but the recreation and other benefits have needed more attention.

COASTAL DEFENCE METHODS

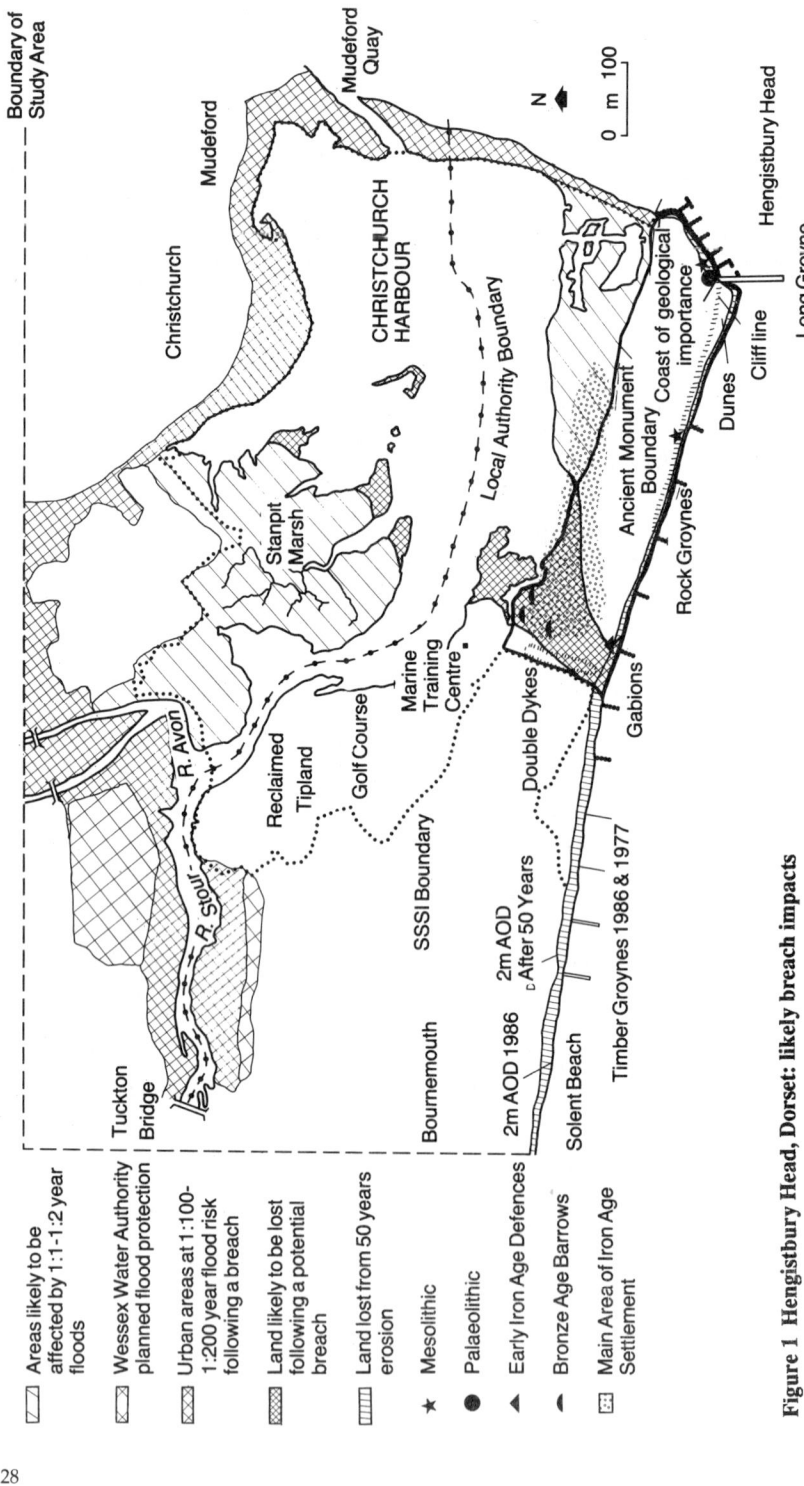

Figure 1 Hengistbury Head, Dorset: likely breach impacts

4. These points are illustrated by three examples, discussed below. The first - Hengistbury Head, Dorset - presents the results of an 'extended' benefit-cost analysis designed to incorporate environmental and recreation benefits into an analysis otherwise dominated by property loss and flood damages. The second example, of the benefit assessment at Fairlight Cove, Sussex (not formally part of the COPRES project) shows the problems that the unknowable future trend of house prices have for the results of benefit assessments. The third example gives some preliminary results from large surveys conducted in 1988 at seven locations in eastern and southern England designed to gauge the benefits of protecting recreational beaches and clifftops from erosion.

HENGISTBURY HEAD: AN 'EXTENDED' BALANCE SHEET APPRAISAL

5. In 1986 an investigation was undertaken with Bournemouth Borough Council of the benefits and costs of coast protection proposals at Hengistbury Head, Dorset (ref. 2). Following this investigation, and parallel physical model analysis of coastal processes and the effects of possible protection works, Bournemouth Borough Council has implemented a scheme comprising a series of timber and rock groynes with beach replenishment. The investigation undertaken by the Flood Hazard Research Centre provides a number of lessons about coast protection appraisal which have general application.

6. Hengistbury Head is connected to Bournemouth by a neck of land which shelters Christchurch Harbour from the prevailing south westerly winds and currents (Fig. 1). The Harbour, which has lower water levels under severe storm conditions than does the open sea, is a sheltered anchorage for a variety of pleasure and fishing craft. Much of the low-lying north shoreline of the harbour is developed for residential and commercial use. The Head is managed as an open space and attracts approximately 740,000 day visitors per annum. The Harbour and Head complex is not only important for recreation. It contains a Local Nature Reserve and was designated in 1986 as a Site of Special Scientific Interest (SSSI) on both environmental and geological grounds. Additionally the Headland contains a rich variety of archaeological sites including the partly excavated Romano-British 'entrepot' which is of international importance.

7. Prior to 1986 the only major coast protection works between Solent Beach and Hengistbury Head was the Long Groyne constructed in the late 1930s. This stretch of coastline was therefore exposed to coastal erosion which is progressing at a rate of about 1.125m per annum in the Double Dykes area - the narrowest and lowest land link between Bournemouth and the Head. To the east of the Long Groyne, cliff erosion is progressing at approximately 0.8m per annum, revealing geological sections of national importance but also causing the loss of ecologically important wind-trimmed heathland.

8. Hydraulics Research Ltd have calculated that a severe storm overtopping the cliffs at Double Dykes could be powerful enough to erode a channel into the Harbour (ref. 2). Such a channel is likely to become a permanent entrance to the Harbour. Several consequences would follow. Land would be lost at Double Dykes and at least part of the entrepot archaeological site would be lost to erosion and the sea. The Harbour would be open to wave action, eliminating or severely reducing the area of safe water, and the tidal regime of the Harbour would be altered such that high tides would be higher and low tides lower. Low-lying properties along the north shoreline of the Harbour, which is already flood-prone, would become prone to more frequent and deeper flooding. Some properties would flood more than once per annum. The Headland would probably become an island, thus limiting access for visitors. The environmental and ecological character of the SSSI would be altered in a complex way and some valuable areas would be lost.

TABLE 1. Extended benefit-cost appraisal 'balance sheet': Wessex Water Authority Scheme; 1.5 mm per year sea level rise; recreational loss valued as boat journey; scheme life 50 years.

Impacts	Without Coast Protection	With Coast Protection
Annual flooding of property		
Number of properties affected by 1:1000 extreme tide	1630	400
PV of flood losses (50 years)	£3.11 million	£0.81 million
Permanent loss		
Property lost (due to frequent flooding with a breach)	61	0
PV property loss (50 years)	£0.85 million	0
Area of open space lost to erosion and a breach	19ha	0
Recreation and harbour use		
Recreational visits hindered per year	740,000	0
PV of lost recreation and extra economic costs	£0.82 million	0
Fraction of harbour uses affected adversely	3/7	0
Ecology and environment		
Ecological diversity (Shannon-Weaver index)	2.11 (if breach)	2.75
Fraction of ecological divisions adversely affected by: erosion	3/11	0
breach	5/11	0
Archaeology		
Fraction of sites adversely affected by erosion	7/8(2/3)[a]	0
Fraction of sites adversely affected by breach	6/8(2/3)[a]	0
Geology: status of Head	nationally important	negligible
Net economic cost to nation	£4.78 million (minimum)	£4.52 million (protection) works and residual flooding

[a] Figure for 'international' status archaeology sites.

9. To reduce the possibility of coastline breaching occurring before a full-scale coast protection scheme could be implemented, Bournemouth Borough Council undertook emergency protection works in 1985 at Double Dykes. Following the results from the physical model study and the economic appraisal, a full scale coast protection scheme was implemented.

The need for an 'extended' appraisal "Balance Sheet"

10. In most cases coastal erosion threatens to destroy environmental assets, but in some cases, such as the gradual exposure of geological strata, erosion is actively creating assets only to 'destroy' them again at a later stage in the erosion process. Similarly coast protection works have both positive and negative environmental benefits and costs, with well designed schemes maximising on the net environmental gains from the necessary investment. There are therefore dangers in utilising narrowly defined economic appraisal methods to evaluate coast protection works. In the worst case those costs and benefits which are considered to be most important to society may be omitted from a narrowly based benefit-cost appraisal.

11. In such cases, an 'extended' economic appraisal method is needed to evaluate coast protection works (ref. 3). An 'extended' economic appraisal is one which involves the identification and quantification of all impacts - whether economic efficiency or environmental - and incorporates these into a single appraisal without necessarily evaluating all impacts in monetary terms.

12. Thus, in the Hengistbury Head study, the benefit-cost analysis not only incorporated potential flood damages and property losses but also estimated environmental and recreational costs and benefits, which are summarised in an 'Extended Benefit-Cost Appraisal Balance Sheet' (Table 1).

The importance of flooding and permanent property loss

13. The Hengistbury Head investigations revealed that the alleviation of property flood damage potential can be an important consequence of alleviating coastal erosion. However, established depth-damage flood damage estimation methods employed in estimating inland riverine flood damage potential (ref. 4; ref. 5) are not transferable to the coast protection case without modification. Employing these data assumes that flood prone property use is unaltered and is recurrent even when flooding frequency increases suddenly as is the case following a breached coastline at Double Dykes.

14. In reality the use of many frequently flooded properties is likely to change substantially - perhaps with ground floors being used less intensively - and some properties may suffer a 'one-off' loss whereby they become economically or socially unusable (ref. 6; ref. 7). For the Hengistbury Head study the dividing line between 'recurrent' flood losses and 'one-off' or permanent property losses was generally taken to be flooding of at least once in two years. 'Recurrent' or 'event' losses are therefore applicable to 393 currently flood-free properties (£214,000), which are expected to become more flood-prone when the coastline is breached. The use is likely to be entirely lost of an additional 61 properties flooded to greater depths by the 2 year flood event or more frequent floods, and these properties were 'written off' with their capitalised values taken as the property loss (£3.15M).

15. In practice little is known about the property use adjustments and changes which are likely to follow from sudden changes in flooding frequency, yet these may be important in cases such as Hengistbury Head and where sea level change may accelerate owing to the 'greenhouse effect'. This is an area where further research is required.

'Externalities' arising from interdependent coastal processes

16. A weakness in the Hengistbury Head appraisal is the omission within the

TABLE 2. Results of 1987 benefit assessment for Fairlight Cove coast protection scheme (costs approximately £1.7M)

	Benefits (£m)
Benefits of protecting residential property (6-year erosion rate; 75 year scheme life; house price differential inflation rate 2 per cent p.a.)	1.31
Westward extension of scheme	0.37
Benefits of protecting undeveloped plots	0.50
Benefits of protecting sewerage system, water supplies and roads	0.19
Total	2.37

TABLE 3. Results of 1988 benefit-cost analysis for Fairlight Cove coast protection scheme (cost approx. £1.72M)

	House Price Differential (% Per Annum)	Discounted Benefits (£m)	Benefit-cost Ratio
1. Erosion profiles based on 7-year average			
	3.0	3.51	2.04
	2.5	3.18	1.84
	1.0	2.47	1.44
	0.0	2.16	1.26
2. Erosion profiles based on 15-year average			
	3.0	3.14	1.82
	2.5	2.81	1.63
	1.0	2.13	1.23
	0.0	1.84	1.07

TABLE 4. Willingness to pay increased rates and taxes to protect the Naze from erosion

	Percent		
	"Yes"	"No"	"Don't Know"
Locals	64	26	10
Day visitors	77	14	10
Staying visitors	76	6	18

'extended' economic appraisal of the evaluation of some of the 'external' impacts beyond the study area of Christchurch Harbour, of coastal erosion and the proposed coast protection works. This is a serious problem which afflicts many project appraisals but which may be very significant where engineering works are proposed for sections of the coast which are interdependent hydrographic and geomorphic units. Whilst the 'extended' economic appraisal took into account the impacts of the proposed river Stour flood alleviation works (Fig. 1), the geomorphic and subsequent economic effects along the coast of continued coastal erosion leading to a breached coastline, and the effects of the coast protection works, were not investigated.

FAIRLIGHT COVE, SUSSEX: BENEFIT ASSESSMENT

17. The village of Fairlight, Sussex, is located at the top of eroding cliffs composed of unconsolidated Hastings Beds. The rate of coastal erosion has increased in recent years, probably owing to the diminished supply of shingle material protecting the base of the cliff, perhaps as a result of harbour and coast protection works further west. Houses and gardens are now very close to the edge of the cliff and, without coast protection works, further losses are inevitable.

18. In 1986 and 1987 Halcrows reported to Rother District Council that protecting the cliff-top houses would be prohibitively expensive (i.e. greater than £4M) for very small benefit (i.e. less than £0.5M). Rother District Council accepted Halcrows' reccomendation not to proceed with a scheme, but suggested that the local community might appraise alternative approaches. The Fairlight Coastal Preservation Association (FCPA) has since prepared two benefit-costs analyses (ref. 8; ref. 9). The first of these benefit assessments (Table 2) produced benefits of £2.37M to compare with a range of less expensive schemes prepared for FCPA by consulting engineers, based on protecting the cliff or involving controlled erosion.

19. However the results in Table 2 contain some assumptions which are problematic. First, they assume that vacant plots in the village are valued as if they could be developed, irrespective of whether planning permission had been granted. They also assumed, more problematically, that house price inflation in the next 75 years (the scheme life) would annually exceed the retail price index inflation by 2 percent (as it has over the last 42 years (ref. 9)).

20. As a result of discussions with the Ministry of Agriculture, Fisheries and Food a revised benefit-cost analysis has been prepared by the FCPA, in conjunction with Middlesex Polytechnic Flood Hazard Research Centre (Table 3). This removes the undeveloped plots without planning permission from the analysis, but improves on the accuracy of benefit calculation and also incorporates a range of relative inflation rates for house prices as sensitivity analyses. Revisions to the benefits of protecting infrastructure may further increase the final benefit figures, but even without these possible increases the results overall would appear to indicate benefits exceeding the reduced anticipated costs of a revised scheme prepared by Dobbies, irrespective of allowing differential house price inflation. Uncertainty remains about the exact current rate of cliff erosion and whether to use the more recent erosion rates (the 7-year average) or a longer time perspective (the 15-year average).

21. The key lessons from this study are that engineers must evaluate low-cost coast protection options as well as conventional structures if they are to meet the benefit-cost test. Also, views on future property price inflation can dominate the calculated benefits, as can the time horizon over which erosion is measured for extrapolating into the future. In locations where houses are more distant from the cliff edge than at Fairlight these considerations will crucially determine whether calculated benefits exceed scheme costs and therefore, as far as the government is concerned, whether coast protection investment is worthwhile.

THE COPRES RECREATION BENEFIT SURVEYS

22. One objective of the Coast Protection Research Project (COPRES) is to develop standard methods and data for gauging the benefits of protecting recreational and amenity land at the coast. However, this is not unproblematic.

23. First, there is surprisingly little previous research on coastal recreation. Therefore initially a broadly based study is required, designed to identify the factors which influence people's choice of a recreational site, their enjoyment of the recreational experience and thence the value they attach to it. Without this data base it was decided to examine local, day and staying or holiday visitors as separate subgroups since it was hypothesised that they would differ in their experience and valuation of coastal recreation.

24. Secondly, selecting study sites from all the coast currently or potentially subject to erosion is problematic. It was decided to concentrate the research on clifftops, beaches, and beaches with promenades, since these are the most common and most intensely used reacreational sites and therefore should generate the most significant coast protection benefits.

25. Thirdly, obtaining reliable and valid measures of recreational benefits is problematic. The Contingent Valuation Method (CVM), widely recognised and used in the United States for evaluating recreational benefits, was adopted for the research (ref. 10). The theoretical and methodological issues involved in applying this method have been reviewed in an earlier paper (ref. 11). The Contingent Valuation Method is a survey method in which money valuations of the recreational benefits of an environmental good such as a beach visit are obtained by direct questioning in personal interviews or through postal questionnaires. Our research also used an indirect method of assessing recreational benefits, the Travel Cost Method developed by Clawson (ref. 12). A major advantage of the Contingent Valuation Method is that it can be used to evaluate future changes in an environmental good such as changes associated with coastal protection or erosion, which is not possible with reference to current travel patterns. Furthermore it can be used to evaluate the recreational benefits of local visitors who incur no travel costs. The results presented below are therefore derived from using the Contingent Valuation Method, using interviews.

Cliff-top recreation survey: Walton on the Naze, Essex

26. A survey of nearly 200 visitors to this site was carried out in October 1987 as a pilot case study of cliff-top recreation. The survey aimed to establish the benefits (in terms of recreational user-benefits) that would be associated with the protection of the cliffs, whilst also testing the methods for measuring these benefits.

27. The problem of continued erosion at the Naze was outlined to the respondents through a statement quantifying the annual rate of erosion. The effect of protection upon the geological character and the general attractiveness of the Naze was presented both through a statement and a photograph of a protection structure similar to the one proposed. A series of questions was asked whereby the respondents were prompted to express their opinion on the likely effects of erosion and protection, and therefore to give careful thought to an issue which may not previously have been considered.

28. The following question was then asked: "Would you be willing to pay increased rates and taxes to protect the Naze from erosion by the sea?". If the response was positive respondents were asked how much more they would be willing to pay. The results showed that some 71 percent of all visitors stated that they would be willing to pay increased rates or taxes to protect the Naze, 18 percent said they would not and 11 percent were undecided. The difference in response in relation to visitor type is shown in Table 4.

29. Average willingness-to-pay figures for each type of visitor, in terms of both monthly and yearly values, were calculated with the assumption that the undecided 11 percent were in fact unwilling to pay increased rates and taxes to provide the protection. Visitors to the Naze are estimated to make around 45,000 trips to the site per year. By grossing up the mean annual willingness-to-pay figures for the different visitor types by their sampled proportions to reflect this level of use, an estimate of the annual benefit of protecting the Naze was derived.

30. Gross annual benefit figures of £85,000 and £108,000 were derived, based on reported monthly and annual willingness to pay respectively. Despite the assumptions inherent in the methodology (and the inclusion of an element of non-user benefit), these figures give a good indication of the likely magnitude of the true benefit figure. With disounting at the 5 percent rate over a scheme life of 50 years this gives a capital sum of some £2M. If coastal protection is carried out to the benefit of all, then it is simply a case of off-setting calculated recreational benefits against scheme construction costs. At the Naze however, a problem occurs in that the geological interest of the site may be damaged by the coast protection works themselves. The costs of "compensating" this user group (which could be evaluated again using the Contingent Valuation Method) should be included in the overall appraisal.

The COPRES 1988 beach recreation surveys

31. Beach recreation surveys have been carried out at three pairs of sites selected to be broadly representative of three different levels and types of recreational development: the large resort site with a high level of commercial development; smaller less commercially developed resort sites; and undeveloped or 'wilderness' sites. One site of each type was chosen in the north east and the south east of England, to explore possible regional differences. Interviews were conducted with beach users on 12 specified days, divided equally between weekdays and Saturdays and Sundays over a period from early July to September. Slightly different versions of the questionnaire were used for local, day and staying visitors and for those interviewed on the promenade at two sites. In addition, visitor counts were taken on four days at each site in order to obtain a more accurate estimate of the proportion of local, day and staying visitors. A total of 737 beach and promenade users were interviewed.

32. The main beach survey was designed to provide answers to a number of key questions, namely how do people value their coastal recreation and what factors are important in determining this value (and thus, ultimately, their willingness-to-pay for coast protection). Additionally the survey was designed to address the problem of valuing enjoyment of recreation, following either coastal erosion or the provision of protection. Therefore, the effect of various changes in the beach environment on enjoyment of recreation and visiting behaviour was examined.

33. *The valuation of recreation.* Respondents were asked to put a monetary value on their "individual enjoyment of this visit to the beach today". To guide them, the respondents were prompted to consider a visit or activity (normally one for which they pay) which gives them approximately the same amount of enjoyment as their visit to the beach. Of the 737 people interviewed at the six sites, 80 percent were able to put a money value on their enjoyment; of those who were not, a similar percentage (80 percent) reported specifically that they "could not value it in money terms".

34. The values of enjoyment given ranged from £ 0.00 to £50.00 with a mean value, across all sites of £7.8 (Table 5). However, there is significant variation in these figures between sites, and analysis is continuing in this respect. The southern sites (Clacton, Frinton and Dunwich) generally show higher mean valuation figures than those in the north (Scarborough, Filey and Spurn Head) although the figures for Spurn Head do not fit this pattern. The average value of enjoyment in the south is £9.20 per visit, whereas in the north the figure is significantly lower (£5.70). Any similarity in valuation that may exist between the pairs of sites (such as was hypo-

COASTAL DEFENCE METHODS

TABLE 5.   Value of "today's enjoyment of the beach visit" (£)

	Mean	Standard deviation	Cases
ALL SITES	7.8	7.1	579
CLACTON	9.9	7.8	153
DUNWICH	6.9	8.2	62
FILEY	3.6	2.2	77
FRINTON	9.5	8.3	125
SCARBOROUGH	5.0	3.8	84
SPURN HEAD	8.5	5.9	78

TABLE 6.   Percentage of all visitors who would get less enjoyment from coastal changes (ranks in brackets)

Changes	Clacton	Scarborough	Dunwich	Spurn Head
Erosion of the prom./dunes/cliff	91 (1)	87 (1)	80 (2)	71 (2)
More crowded beach	85 (2)	76 (2)	96 (1)	91 (1)
Less sand, more pebbles	80 (3)	74 (3)	42	40 (3)
More groynes/breakwaters	70 (4)	33	48 (4)	20
Steeper beach between high and low tide	69 (5)	26	27	9
No beach above high tide	63	40 (4)	53 (3)	30 (4)
Steeper beach above high tide	68	34 (5)	43 (5)	19
Beach narrower at low tide and uncovered for less time	68	27	39	25 (5)

thesised in our site selection) has been over-riden by other factors including, it seems, those differences in socio-economic factors summarised by geographical location. Indeed, real and pronounced differences are evident in the income distribution of respondents in the north and south although Spurn Head stands out as atypical of the general picture in the north.

35. If the differences in valuation of enjoyment are to be explained more fully, a factor that must be looked at is the importance of the beach, as opposed to other resort characteristics and considerations, in drawing visitors to a site. Furthermore, differences in the type of "experience" offered by each of the six sites may affect the types of visitors attracted and therefore the recreational value.

36. *Differences in recreational 'experience'*. Significantly, the factor that emerges as the most important reason for the respondents' choice of recreational site is that they "wanted to visit the coast". This immediately identifies the distinction between coastal and non-coastal recreation. Also of importance to the visitor population as a whole is the attractive and natural setting of the beach and its "cleanliness".

37. The differing importance of other factors in attracting the visitors illuminates significant differences in the type of recreational experience at each site. For example, at Scarborough, the additional attractors that emerge strongly are that the visitors have "liked it when they have been before" and that the "beach is good for children". No doubt allied to its suitability for children, the visitors cite as very important attractors the fact that the beach is "spacious" and "sandy", and that it has good associated facilities. In contrast, at both Spurn Head and Dunwich, as was hypothesised, the key attractors relate to the attractive natural environment and quietness.

38. Just as different factors have been important in attracting visitors to the different beaches, so these beaches have been shown to exhibit different patterns of use. At Clacton, for example, the visitors interviewed reportedly undertake a number of activities (on average over 4 each), the most common of these being short walks along the promenade (85 percent of visitors), sitting/picnicking on the beach (67 percent), short walks along the beach (61 percent), sitting/picnicking on the promenade (58 percent) and swimming/paddling (52 percent). At Spurn Head, however, the only activity undertaken by a large proportion of the visitor population is short walks along the beach (76 percent). Short walks along the dunes and long walks on the beach and dunes were of some importance (23 percent each) but other activities commonly undertaken at Clacton such as sitting on the beach (14 percent) and swimming/paddling (10 percent) were relatively insignificant.

39. *The importance of beach characteristics*. An exploration of the data indicates that not only do the broader characteristics of the location affect the activities undertaken on the beaches (by influencing the types of users that are attracted) but the physical characteristics of the beach itself are also very important. For example the pebbly beach at Dunwich was reported to be uncomfortable under foot and thus only some 4 percent of visitors reported that they went for long walks, as opposed to 25 percent at both Spurn Head and Clacton.

40. The importance to different users/activities of factors such as the slope and profile of the beach and the amount of beach material remaining above high tide is less immediately clear from the results. This may be because respondents find it difficult to disaggregate the beach into its various physical features. Preliminary results do, however, indicate that the physical characteristics of the beach are quite strongly linked to functionality and therefore they are of importance in determining recreational behaviour.

41. The example of the effect of groynes/breakwaters is well illustrated at Clacton. Overall the beach at Clacton is believed (by the majority of visitors) to be "good" for each of the activities of walking, sitting/lying and for children to play on. The presence of groynes/breakwaters is seen as being of importance in making the beach

TABLE 7. The effect of beach erosion on visitors' enjoyment and visit frequency

	Visitors who would get less enjoyment from their visits (percent)	*Visitors who would visit the beach less often as a result (percent)*
CLACTON	55	30
DUNWICH	43	20
FILEY	57	30
FRINTON	53	28
SCARBOROUGH	73	69
SPURN HEAD	42	19

TABLE 8. Hastings seafront users survey: first stage results

A. Values for beach enjoyment based on drawings of alternative scenarios: mean values (£)

	Overall	*Locals*	*Day Visitors*	*Staying Visitors*
"Today's visit"	7.7	6.0	9.9	6.6
Existing beach drawing	6.6	7.8	6.1	6.2
Eroded beach drawing	3.8	3.8	2.8	5.0
Protected beach drawing	11.5	12.6	10.6	11.6

B. Willingness-to-pay for coast protection via "increased rates and taxes"

Percent willing-to-pay	77	65	84	81

good for children (by restricting their wandering) and for sitting/lying (by providing shelter from the wind) but relatively unimportant in making the beach good for walking. Indeed, for the 10 percent of visitors who believe Clacton beach is "poor" for walking on, the presence of breakwaters and groynes is cited as a contributing factor by the vast majority (88 percent) of these visitors.

42. *The effect of erosion.* Having established that the "functionality" of the beaches (and also the beach characteristics required for some of these activities) varies, it was hypothesised that due to these differences, any changes in the physical characteristics and usage of the beach (such as may be associated with either beach erosion or protection) would affect visitors differently at the different sites.

43. At both Scarborough and Clacton, where the beaches have been identified as good for children and for playing games etc, results indicate that if the beaches became less sandy and more pebbly this would reduce the enjoyment of the majority of the visitors (75-80 percent)(Table 6). At Dunwich and Spurn Head this change will adversely affect a far smaller proportion of visitors (40 percent). An increase in crowding of the beach is the change seen as having by far the greatest adverse effect at Dunwich and Spurn Head, and this highlights their function in providing a 'wilderness experience'. Interestingly, at Spurn Head changes in the individual physical characteristics of the beach (including the presence of more groynes) do not greatly affect the enjoyment of the visitors. However, at Clacton, Dunwich and Scarborough, if the beach became steeper and totally covered at high tide, or indeed had more groynes, the percentages of the visitors deriving less enjoyment from a visit to the beach are significantly higher than for Spurn Head.

44. The potential effect of erosion on people's enjoyment and frequency of visits to the beaches was gauged from the following questions:

> "With coastal erosion, beaches usually become narrower. If this beach became narrower at high tide, would you get more or less enjoyment from your visits to it or would it make no difference (exact question wording varied with sites)?"

> "If this change were to occur, would you visit the beach more often, less often or as often as now?"

Although the questions asked were quite general, the responses derived give some idea of the effects of erosion on beach recreation, hence the potential importance of recreational benefits for justifying coast protection works.

45. Overall, 54 percent of respondents indicated that their enjoyment would be diminished if the reported erosive change occurred, and 32 percent of the respondents would visit the beach less often than now. The fact that only 2 percent and 6 percent respectively of visitors were unable to say how the erosion would affect their enjoyment and visiting frequency, highlights the ability of the user population as a whole to perceive physical changes and to relate these to their recreational behaviour.

46. As Table 7 shows, the likely effect of the erosive change varies greatly between sites. Visitors to Scarborough can be seen to be most "sensitive" to the changes in the beach described, with a large percentage of visitors (69 percent) stating that they would visit Scarborough beach less often as a result of the erosion. At Spurn Head and Dunwich, although a significant proportion of the visitor population would be adversely affected by the erosion, the proportions are far less than for Scarborough and the other sites. Indeed only 19 and 20 percent, respectively, of visitors indicate that they would visit Spurn Head and Dunwich less often as a result of the possible changes.

47. Whilst there are always problems with this type of CVM research, there is nevertheless unquestionable evidence that the effect of any given erosive change upon the recreational users of beaches is related to the prevalant functionality of the beach. This is because the functionality is directly related to the physical characteristics of

COASTAL DEFENCE METHODS

the beach, as well as to the broader characteristics of the location. The less pronounced effect at Spurn Head than at Scarborough of essentially the same potential change to the beach, reflects the centrality and importance of the beach to the recreational experience at Scarborough, as opposed to the greater importance of isolation and naturalness at Spurn Head.

THE HASTINGS, SUSSEX, SEAFRONT USER SURVEY

48. A coast protection scheme is being installed in stages at Hastings, Sussex, and a survey of 247 beach and promenade users was carried out during August and September 1988. The focus was on users' perceptions of specific beach changes resulting from coastal erosion and protection. Respondents were asked both to suggest a monetary value for the enjoyment of their visit and use of the beach, and to indicate their willingness-to-pay for sea defence and coast protection schemes through "increased rates and taxes".

49. In this research, in contrast to the studies described above, drawings were used to portray these potential beach changes, based on design specifications provided by the Hastings engineers. Three beach scenarios were thus compared: current conditions; the beach following increased erosion without coast protection works; and the beach following the completion of the coast protection works. The drawings were based on a section of beach which is to be protected during the later phases of the scheme, and a repeat survey is anticipated after the full scheme is completed.

50. Respondents' ability to understand quite complex questions requiring them to assess and evaluate different beach characteristics was validated by comparing responses from sections of the beach showing different characteristics. Amongst the features evaluated were the condition of the groynes, breakwaters, the sea wall and promenade.

51. To enable comparisons to be made between the different scenarios, respondents were asked to value their enjoyment of their visit and use of the beach. Overall, four percent of respondents were unable to answer this question and 14 percent were unwilling to value their enjoyment in monetary terms. There were, however, significant differences in responses from local people and other visitors, which may be related to age differences between these user groups.

52. The survey results for the valuation of enjoyment of the beach in its current condition gave a mean value of £7.7 per visit (Table 8). On average, all user groups consistently gave the eroded beach lower values and the protected beach higher values than the existing beach. The locals valued the existing beach and the protected beach more highly than did other visitors, although they were less willing to pay for coast protection. This can be related to respondents' age, with a higher proportion of locals than visitors in the 65+ category and a lesser ability to pay increased rates evident in this age group. Overall, the results indicated that the public are well able to distinguish between different beach conditions and to put a monetary value on their enjoyment of them in accordance with clear preferences.

CONCLUSIONS

53. The research reported in this paper is continuing and the results from the COPRES project are preliminary. Nevertheless it is apparent from both the Hengistbury and the Fairlight examples that coast protection will be difficult to justify in narrow benefit-cost terms unless the circumstances are exceptional. In the Hengistbury case the exeptional character is the possibility of a breach, and the consequent damage from flooding and permanent property loss. In the Fairlight case the benefits are a function of the nearness of the cliff edge to the property at risk, and thus the immediacy of the hazard that the community faces.

54. The likely problems with narrow benefit-cost analysis means that the appraisal must be conditioned by two parallel initiatives. First, we must look critically at

scheme design and cost, and evaluate low cost options even if they do not 'solve' all of the coast protection and management problems at the site in question (ie there are residual losses that have to be accepted). Secondly, it is essential that the appraisal is extended to evaluate recreational and amenity benefits. Our research at the Naze, the six main COPRES sites, and at Hastings has shown these to be substantial. In addition we must evaluate ecological, geological and other important benefit categories, to ensure that our "balance sheets" for comparing 'with' and 'without' scheme scenarios are complete.

55. Only in this way will coastal protection and management be comprehensive and systematic. In addition, a broad approach to project appraisal will mean that the nation's growing emphasis on environmental protection and nature conservation can be adequately incorporated into investment decisions.

REFERENCES

1. THOMPSON P.M., PENNING-ROWSELL E.C., PARKER D.J. and HILL M.I. Interim guidelines for the economic evaluation of coast protection and sea defence schemes. Middlesex Polytechnic Flood Hazard Research Centre, London, 1987.

2. HYDRAULICS RESEARCH Ltd. Hengistbury Head coast protection study. Hydraulics Research Ltd, Wallingford, 1986, Summary and Technical Reports EX1459 and EX1460.

3. PARKER D.J. and THOMPSON P.M. An 'extended' economic appraisal of coast protection works: a case study of Hengistbury Head, England. Ocean and Shoreline Management, 1988, vol 11, 45-72.

4. PARKER D.J., GREEN C.H., and THOMPSON P.M. Urban flood protection benefits: a project appraisal guide. Gower Technical Press, Aldershot, 1987.

5. PENNING-ROWSELL E.C. and CHATTERTON J.B. The benefits of flood alleviation: a manual of assessment techniques. Gower Technical Press, Aldershot, 1977.

6. THOMPSON P.M. and PARKER D.J., Hengistbury Head coast protection proposals: assessment of potential benefits and costs, Summary and Supplementary Reports. Middlesex Polytechnic Flood Hazard Research Centre, London, 1986.

7. THOMPSON P.M., PARKER D.J., COKER A., GRANT E., PENNING-ROWSELL E.C. and SULEMAN M. The economic and environmental impacts of coast erosion and protection: a case study of Hengistbury Head and Christchurch Harbour, England. Middlesex Polytechnic, London, 1987, Geography and Planning Paper.19.

8. FAIRLIGHT COASTAL PRESERVATION ASSOCIATION. Cost/benefit analysis. F.C.P.A., Sussex, 1987.

9. FAIRLIGHT COASTAL PRESERVATION ASSOCIATION. Fairlight Cove, Sussex: benefit-cost analysis. F.C.P.A., Sussex, 1988.

10. CUMMINGS R.G., BROOKSHIRE D.S. and SCHULZE W.D. Valuing environmental goods: an assessment of the Contingent Valuation Method. Totoya: Roman and Allanheld, New Jersey, 1986.

11. TUNSTALL S.M., GREEN C.H. and LORD, J. The contingent valuation method. Middlesex Polytechnic Flood Hazard Research Centre, London, 1988.

12. CLAWSON M. Methods of measuring the demand for outdoor recreation. Washington: Resources for the Future, 10, 1959.

20. Case study at Carmarthen Bay

P. C. BARBER, MEng, PhD, MICE, MBIM, Managing Director, BMT Ceemaid Ltd, and R. P. THOMAS, MSc, MICE, Deputy Borough Engineer, Llanelli Borough Council

The paper describes an approach to coastline management related to a coastal 'cell' defined by consideration of the environmental processes appertaining. The cell comprises Carmarthen Bay and the approach has been sponsored by six Public Authorites with frontage interests within the bay area.

A study has been commissioned to assemble and collate the relevant data base; to supplement the data base and to interpret the results of this work set against consortium objectives.

The study described here was carried out in three stages commencing in 1985 and completed in January '89. The study outcome has provided each sponsor with specific definition of the environment and design criteria for coastal works upon his frontage together with a strategic definition of how the larger scale processes acting in the bay affect his interests.

Development of this approach to coastline management is now set to provide significant savings; amenity benefits and balanced conservation during the future programme of coastal works in the Carmarthen Bay area.

OVERALL AIMS AND APPROACH TO STUDY

Introduction

1. The geographic area identified as Carmarthen Bay has a coastline passing through 4 Local Authority areas, from Swansea in the East, through Llanelli and Carmarthen to South Pembrokeshire in the West.

2. It is made up of varying features including the wide flat dune-backed beaches of Pendine and Pembrey; the Taf, Tywi, Gwendraeth Estuary complex; the Burry Estuary and the hard rocky promontories of Western Gower and South Pembrokeshire.

3. During the early 19th century the towns of Swansea and Llanelli witnessed dramatic changes in their industrial development. Although the growth of Swansea had little effect

ENGINEERING STUDIES

on Carmarthen Bay, the docks and industrial sites being somewhat removed from the area, the industrialisation of Llanelli saw the construction of the docks, a coastline railway and a channel training wall which had a significant effect on the Burry Inlet and consequently Carmarthen Bay.

4. Much of the western half of the bay had always been associated with tourism, as had the extreme eastern boundary, being part of the Gower Peninsula.

5. However, the steady decline in heavy engineering, which commenced in the 1950's culminating in the closure of the Dupont Steelworks in the early 1980's lead Llanelli Borough Council to realise the need to remove what industrial dereliction remained and to formulate a strategy for developing its coastline for recreation and tourism.

6. Besides the Local Authorities, responsibility for coast protection in Carmarthen Bay lay with three organisations: British Rail, protecting coastline railway embankments, the P.S.A., acting as agents for the Ministry of Defence's ranges at Pendine and Pembrey and Welsh Water Authority. Until the mid 1980's these seven bodies had undertaken their responsibilities for the coastline individually without any reference to each other and without full understanding of the forces acting within the Bay as a whole. In other words, the coastal engineering problems tended to be treated symptomatically rather than fundamentally.

7. In 1985 it was realised that similar problems were being experienced by the various organisations and a consortium of six of those most affected was established to study the behaviour of the tidal regime of the Bay.

8. The Consortium comprised: Carmarthen District Council, British Rail, Swansea City Council, P.S.A., Welsh Water Authority and Llanelli Borough Council. From the Group's inauguration the Welsh Office gave its full support, welcoming the idea of unified approach to the study of coast protection: The Polytechnic of Wales was also involved in the Group from the outset in an independent advisory capacity.

9. It was obvious that, for the greatest benefit to be gained from an examination of the Bay's behaviour, a specialist consultant would need to be employed to study the problems. Consequently, it was agreed that a number of specialists would be requested to put forward their proposals for undertaking a study, dividing the work into discrete phases, each of which could stand alone as a seperate report.

10. The selection procedure involved the consideration of several proposals and, following the appointment of the successful consultant, British Maritime Technology, the first

phase was commenced, involving the collection, examination, and correlation of all existing available data.

Study Rationale

11. The understanding of a section of shoreline is typically defined in U.K. by local problem or more generally between administrative boundaries. Present knowledge of the natural processes governing shoreline behaviour seldom accords with such definitions of problem zone.
In addition resources are typically limited and the engineer must address the matter of providing sound advice on problem solution within restricted budgets. The use of numerical modelling techniques to assist in the development of solution strategies to such problems provides an attractive modern option but it is essential that a balance is maintained between study objectives, client budget and the technique applied.

12. It is important to address the following in the formulation of study definition:

- available resources,
- client objectives,
- principal natural processes appertaining,
- appropriate techniques (numerical, physical models, measurement, analysis etc),
- system of technique validation
- design parameters to be established and their accuracy,
- economic assessments of solution options.

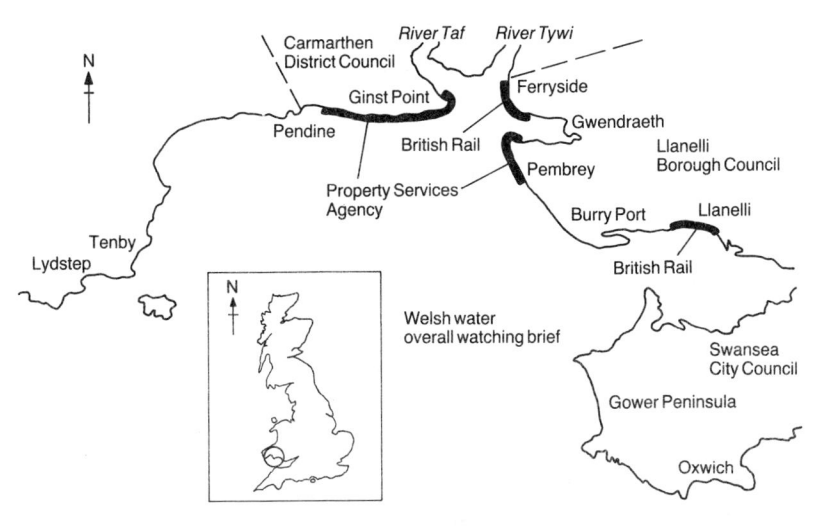

Fig. 1. Map of Carmarthen Bay showing location of consortium member boundaries; Associated Member: Polytechnic of Wales

ENGINEERING STUDIES

These seven 'pillars of wisdom' need to be defined before a study is commenced.

13. At Carmarthen Bay, six Authorities responsible for the coast protection (figure 1) have in the past identified problems relating to hydrodynamics of the region and have tended to take action in isolation to the specific site without an understanding of the more wide reaching effects within the bay as a whole.

14. The main objective of the study has thus been to obtain a quantitative and qualitative understanding of environmental processes within Carmarthen Bay. This understanding will lead to cost effective management of the coastal resource and in particular will assist in:-

a) The more efficient discharge of local studies for individual developments.

b) Joint schemes between Authorities.

c) The effect of individual schemes on adjacent sections of the coastline to be elevated.

d) Establishment of design parameters for areas where immediate remedial work is required.

e) Prediction of future areas of concern.

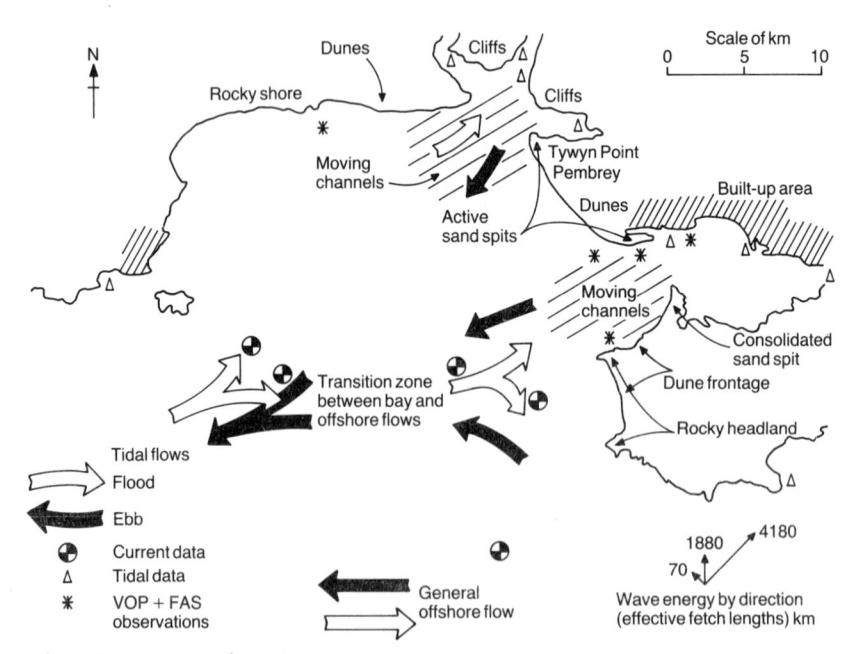

Fig. 2. Synoptic plan of Carmarthen Bay

Approach to Study

15. Carmarthen Bay extends some 42km east to west and indents some 24km. The coastline covered by the study extends approximately 70km and varies from extensive sand beaches to mud flats or rocky cliffs (figure 2). Like wise usage of the coastal strip varies from industrial to recreational, agricultural to residential. It was evident early in the definition of the study content that numerical modelling provided a major capability in the achievement of study objectives. It was important to examine a competent balance of of study techniques.

Table 1 Field Measurements

	Location	Dates of Measurement From To
Tidal Data		
Tidal Elevations	L. O, B	17.7.87 12.8.87
Tidal Dips	Estuaries	15.7.87 15.7.87
Current Metering	W, I	(W)17.7.87 12.8.87
	(E)	25.7.87
Float Tracking	Burry Inlet	12.8.87 13.8.87
Admiralty diamonds	E, F, G	1894 1947
Wave Data		
Fixed Aspect Stills	Various	Winter 1985/6 and winter 1987/8
Video Records	Various	Winter 1985/6 and winter 1987/8
Wave Measurements Offshore (IOS)	St. Gowan's Helwick	1975 1978 Aug 1960 July 61
Sediment Data		
Sediment Traps	Pendine (Tide only)	3.7.87 7.8.87
	Pendine (Tide & wave)	Winter 1987/8
	Pembrey (Tide only)	Winter 1987/8
	Pembrey (Tide & wave)	Winter 1987/8

ENGINEERING STUDIES

16. The primary natural agents applying to the Bay area required a numerical model to include the effects of waves and currents and their interaction. Restrictions on resources required the development of a careful approach to model validation. It was decided to undertake specific flow validation but site limitations of wave validation lead to the adoption of shoreline measurements with model validation undertaken by correlation techniques. In this it was recognised that numerical wave models were not easily assembled to predict shoreline behaviour but that this was critical to coastline behaviour, an essential study objective. The adoption of this strategy has resulted in the successful integration of available resources. A list of measurements undertaken in the project, and previous information available is given in Table 1.

The study has comprised three stages:-

a) Data Base collation and establishment of monitoring

b) Numerical modelling and field measurement validation

c) Model application and definition of regime and design criteria.

Monitoring has been undertaken by the Polytechnic of Wales and permitted the cost effective integration of local resources with more distant specialist expertise. The various monitoring techniques are described further in paragraphs 31-37.

AVAILABLE RESOURCES

17. The initial proposal for the study recommended the three-stage approach with an overall study budget of £100,000 (1985 price-base). The division of the study into stages allowed milestone reviews of progress and future direction. Although the technical approach has altered as data were assessed it has been possible to maintain costs within the original budget ceiling (inflation allowance notwithstanding) and achieve client objectives.

CLIENT OBJECTIVES

a) To obtain a definition of natural processes affecting the study area with specific regard to the shoreline influence.

b) To relate the definition from 1 by the establishment of design criteria to particular problem areas and to adduce associated solution strategies.

c) To identify strategic policy options of coastline management within the Bay area extending across administrative boundaries with definition of costs and benefits to Consortium members.

d) To recommend future actions of the Consortium to exploit and develop study results.

NEARSHORE PROCESSES

18. The study area is shown in figure 2 which also shows the synoptic definition of the major natural processes. The bay shape is dominated by the high energy south-westerly wave approach, interrupted by the flushing of the two major estuary systems discharging into the bay within the high tidal range.

Tidal Currents

19. The tidal range is approximately 8m on a mean spring tide. This gives rise to maximum tidal currents in excess of 1m/s in the main bay area. Tidal current action has three dominant effects:

- a) Effects on waves entering the bay. This results in significant current refraction on waves particularly on the ebb tide.
- b) Direct scour action at particular coastal sites, for example at Ginst and Tywyn Points.
- c) Providing transport mechanisms to move sediment stirred by wave action.

20. The outlet of three major rivers into the bay, in combination with estuarial tidal prisms, has resulted in a complex system of bars and channels being formed at the estuary mouths. This has a substantial effect on wave conditions in these areas.

Wave Action

21. The major shallow water processes (Barber 1985) affecting waves as they travel inshore are as follows:

- a) Depth refraction, particularly over the shallow inlet areas.
- b) Current refraction.
- c) Partial wave breaking over the shallow inlets and full wave breaking at the shore.
- d) Bottom friction and shoaling, especially over the wide sandy foreshores off Pembrey and Pendine Sands.

Wind Action

22. Wind action has been identified as a significant mover of material from the large expanses of intertidal banks. The harnessing and constructive use of such material movement is also an important aim of the study.

ENGINEERING STUDIES

To this end and also to establish typical wave climates a hindcasting wind/wave model has been run using wind data over a ten year (1974-1983) period.

MODEL DESCRIPTION

Tide Model

23. The tide model is based on the depth integrated 2-dimensional equations of motion, Viz:

$$\frac{\partial \eta}{\partial t} + \frac{\partial Ud}{\partial x} + \frac{\partial Vd}{\partial y} = 0 \tag{1}$$

$$\frac{\partial U}{\partial t} + U\frac{\partial U}{\partial x} + V\frac{\partial U}{\partial y} - fV + \frac{kU}{d}\sqrt{U^2+V^2} + g\frac{\partial \eta}{\partial x} = 0 \tag{2}$$

$$\frac{\partial V}{\partial t} + V\frac{\partial V}{\partial y} + U\frac{\partial V}{\partial x} + fU + \frac{kV}{d}\sqrt{U^2+V^2} + g\frac{\partial \eta}{\partial y} = 0 \tag{3}$$

where

η = mean surface elevation (relative to M.S.L.) in metres.

t = time (s)

U = x component of depth mean current velocity in m/s.

V = y component of depth mean current velocity in m/s.

x = x axis in cartesian coordinates in metres.

y = y axis in cartesian coordinates in metres.

F = Coriolis coefficient.

k = friction coefficient.

d = water depth in metres.

g = acceleration due to gravity in m²/s.

24. The model uses the Alternating Direction (or Angled Derivative) Explicit (ADE) scheme, of Flather and Heaps (1975), but modifies their method to incorporate a two stage approach for the calculation of U and V velocities (Yoo and Hedges,1985). This results in increased stability of the model over steeply sloping bathymetry.

25. One major need was for a nesting approach whereby local areas of coast could be studied in fine detail without losing definition offshore. This facilitated the assessment of Hydrodynamic conditions in the vicinity of coastal structures, and also their effect on the conditions in the rest of the bay.

26. Coarse (M1), medium (M2), and fine (M3) meshes have been set up to cover the offshore area (from Lydstep to Oxwich Bay and 9km offshore from Helwick Bank) at a spacing of 2km, nearshore areas at a spacing of 666m (see figure 3) and areas of rapid bathymetric and flow change (i.e. both estuaries) at a spacing of 222m.

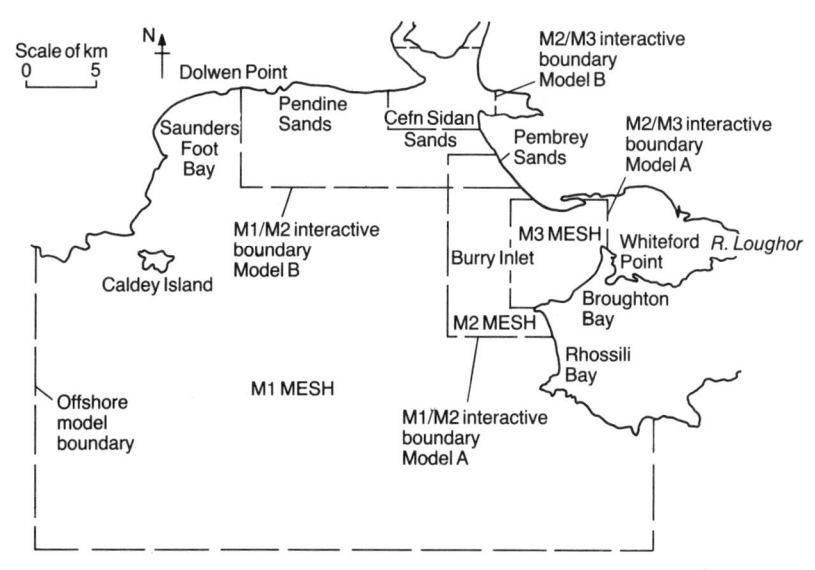

Fig. 3. Model grid layout for Burry Inlet and Cefn Sidan Sands: Carmarthen Bay flow model M1

A wetting and drying scheme is used to model flows over drying banks.

Wave Model

27. The wave model is based on the kinematic and dynamic conservation equations:

$$\frac{dK_i}{dt} + K_j \frac{\partial U_j}{\partial x_i} + S_h \frac{\partial d}{\partial x_i} - \frac{R_i}{2Ka} \frac{\partial^3 a}{\partial X_i^2 \partial X_j} = 0 \qquad (4)$$

$$\frac{da}{dt} + \frac{a}{2} \frac{\partial}{\partial x_i}(R_i + U_i) + \frac{a}{2} X_{ij} \frac{\partial U_i}{\partial x_i} + \frac{\varepsilon^*}{\rho g a} = 0 \qquad (5)$$

where

K_i = wave number vector K = wave number
σ_o = intrinsic angular frequency S_h = depth variation factor $\partial \sigma_o / \partial d$
a = wave amplitude
R_i = wave group velocity
X_{ij} = $\delta S_{ij}/\delta E(K_i)$
δS_{ij} = radiation stresses of component waves in spectrum
δE = wave energy density of component waves in spectrum
ε^* = enhanced wave energy dissipation rate
ρ = water density

where i = 1,2 (as in conventional index notation)

ENGINEERING STUDIES

28. The equations on the previous page are supplemented by Doppler equation and dispersion relationship for linear waves. The modified equations are then solved along the wave rays by the method of characteristics. Diffraction is accounted for by a third order term in the kinematic conservation equation. Checks for the complete and partial breaking of waves are also made.

29. The bottom friction model for combined flow was developed by Yoo (1986) who extended and refined Bijker's (1967) to include the effects of current reduction and the increase of wave boundary layer thickness.

30. The wave model uses the same mesh systems as the tidal model (M1, M2, M3). Outout of wave amplitude and direction, although initially available along wave rays, is also available on the regular meshes, to aid interpretation of conditions nearshore. Combined wave and tidal energy flux vectors are also output from which an interpretation of sediment movement can be made.

MONITORING AND MODEL VALIDATION

Tidal Model Validation Measurement

31. The model was run initially for a MHWS tidal range using harmonic constituents of tidal amplitude and phase derived from a larger model of the Bristol Channel (Owen 1980). This was representative offshore input data but required verification against measured tidal data along the offshore boundary. These were obtained at the east and west boundaries of the model (Lydstep and Oxwich Bays). An additional reason for deploying gauges at these points was to establish a time base on which other measurement within the model might be judged.

32. Two long term current meters were deployed between Caldey Island and the Burry Inlet (see figure 2). This together with source data from Admiralty diamonds provided a good control on the interaction between flows into and out of the bay and the east-west flow in the Bristol Channel.

33. It was also important to establish tide level validation within Taf/Tywi and Loughor (Burry Inlet) estuaries. This was carried out by taking dipped records from slipways, bridges, jetties etc, at seven locations. Although marginally less accurate, the coverage was far more than could have been achieved through deploying tide gauges on the same budget.

34. Float tracking was also carried out in the Burry Inlet and in the Taf/Tywi estuary.

35. The initial runs of the model used latest available bathymetry obtained by the Hydrographer to the Navy. These

(a)

(b)

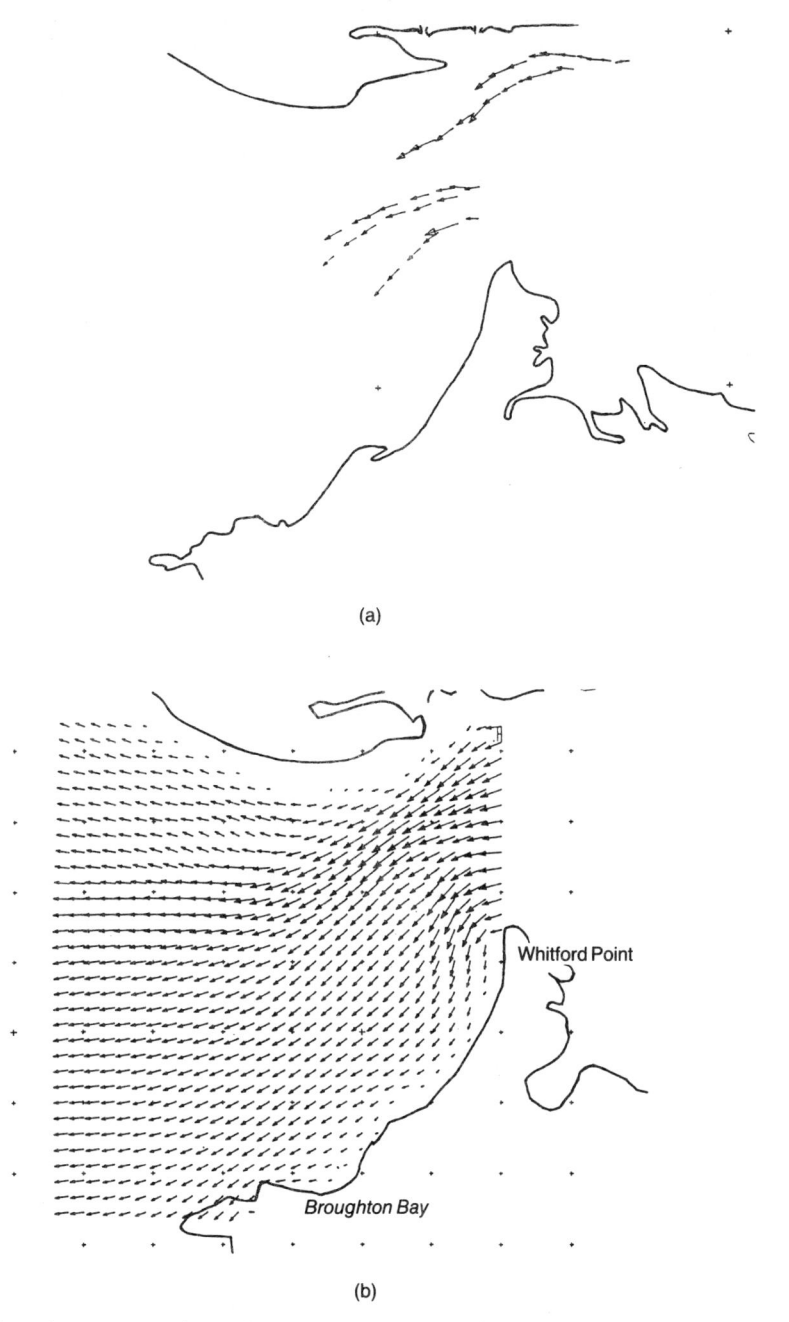

Fig. 4. Comparison between measured and predicted flow, M3 grid area Burry Inlet, HW -2 hours: (a) field data; (b) model data

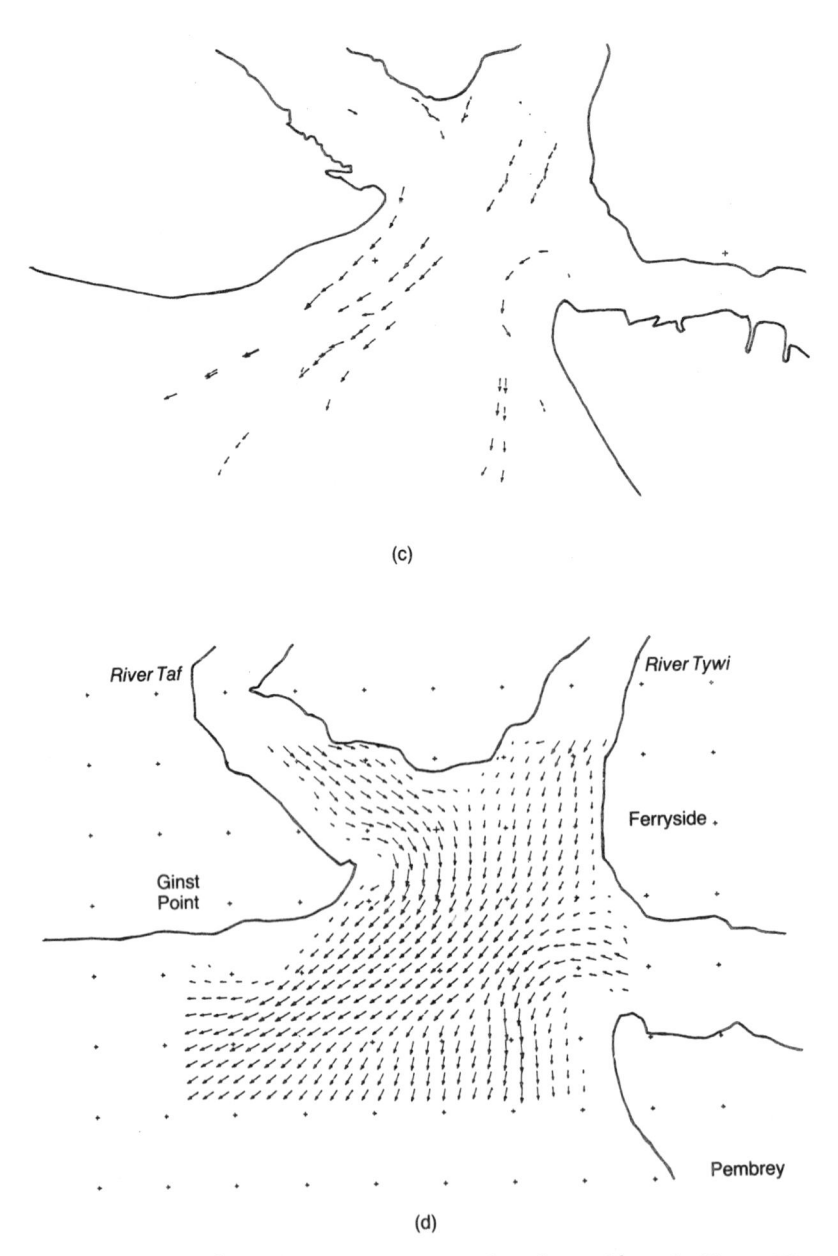

Fig. 4. Comparison between measured and predicted flow, M3 grid area Northern Estuary, HW +2 hours: (c) field data; (d) model data

levels dated back to 1977 in the Taf/Tywi and Loughor/Burry Inlet estuaries. Following these original model runs it was apparent that the changes in estuary morphology identified in the collated data base would be important in establishing not only present flow conditions but also future trends. A coarse grid bathymetric survey was therefore undertaken, the results of which highlighted major reconfiguration in channels and banks.

36. Comparison of model results against current measurement offshore and against float track measurements within both estuary systems showed good agreement. Figures 4a and b (Burry Inlet) and figures 4c and d (Taf/Tywi) present comparisons of tidal current predictions. Tidal elevation predictions at the dip sites also produced good agreement.

Wave and Sediment Validation

37. It is more difficult to measure nearshore wave and sediment movement than deep water circulations. Whereas the tidal movements occur with a predicted regularity and usually sufficient calm weather windows, rough weather of interest, inshore, occurs sporadically. The mounting of a detailed measurement programme within the intertidal zone can often produce useful data when required at a specific location. However the requirement was for long streches of coastline to be covered without prohibitive costs. To this end, three simple monitoring techniques were used:-

   a) Visually observed poles. (VOP)

Graduated poles were erected at six locations around the bay (see figure 2). Video records were then taken by local personnel during storm conditions. These records were then analysed for wave height (Hs) and period. Further information on wave groups, breaking waves and surf zone width may also be obtained to help validate wave model results.

   b) Fixed Aspect Stills (FAS)

This monitoring, normally carried out in association with the VOP-video work. provides a definition of inshore wave direction. Oblique photographs are taken of the sea states from surveyed locations at repeatable horizontal and vertical orientations (using land features to fix the picture). From these, an oblique grid may be constructed over the photographic image. It is then a relatively simple task to identify wave fronts and relate them to a normal plan grid for direction definition. Figure 5 shows a typical plan view of wave crests obtained in this way; 5b shows the model prediction plot for a similar wave condition.

   c) Sediment Traps

These traps consist of 2.5m long tubes of filter cloth closed at one end and supported from a scaffold tube frame (see Kraus, 1987). They were deployed at three locations along the Pendine

ENGINEERING STUDIES

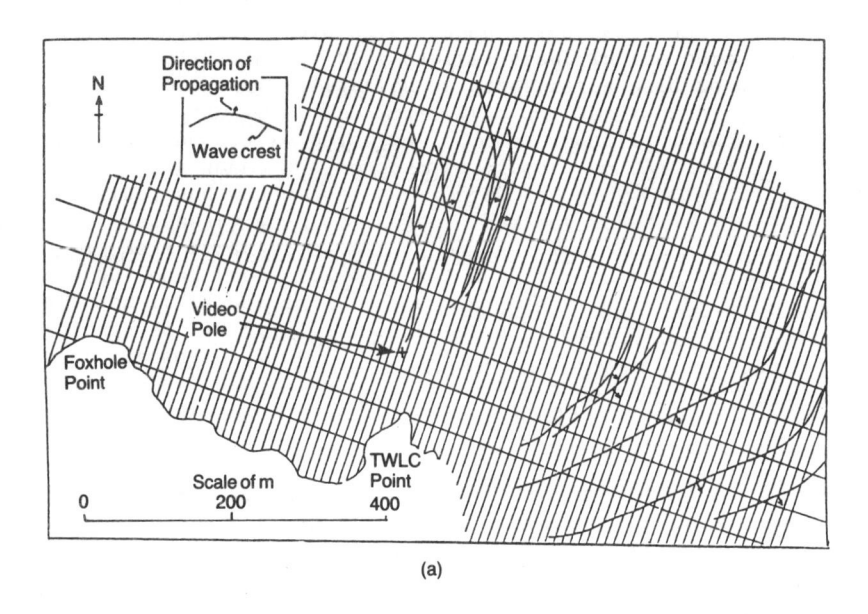

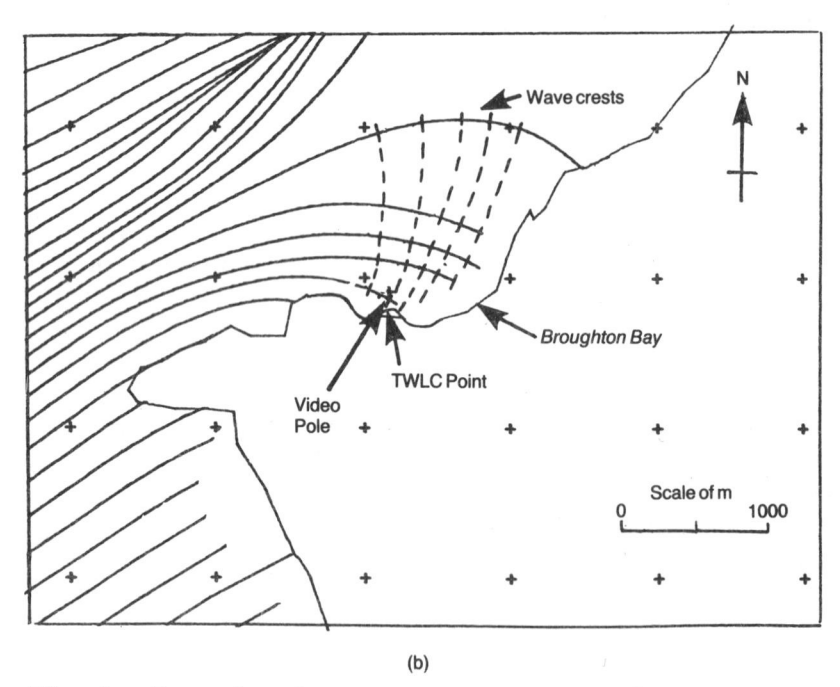

Fig. 5. Comparison between measured and predicted wave conditions at Broughton Bay: (a) field data, per = 9.0 s; (b) model prediction, per = 8.0 s

frontage orientated parallel to the coast (in two directions at each locations) and left for seven day periods over calm weather during the summer. Significant variation was noted in the amount of material collected between each site and between the two traps at each location. Traps were also deployed in the same locations over individual storm tides during the winter months of 1987/88.

ADVANTAGES AND LIMITATIONS OF TECHNIQUES USED

Advantages

38. The major benefits and limitations of the techniques applied to the bay are given below.
   a) The model has been set up as an aid to understanding the overall hydrodynamics of the bay system.
   b) All major tidal and wave effects are modelled.
   c) The model has a nesting facility, which means that particular sections of interest can be modelled accurately without incurring overhead penalties for the larger meshes.
   d) The low cost techniques of monitoring provide good spatial coverage, as well as providing an on-site visual record of actual processes.
   e) The model has been set up for bathymetry measured in 1977 and 1988 (part revision). This may easily be modified to take in changes in the bathymetry in order to study their hydrodynamic effects.

Limitations
   a) Linear wave theory is used in the model which is not valid inside the surf zone. The results of the present model are to be interrupted here using data from wave monitoring.
   b) The wave model considers only monochromatic unidirectional waves. The video techniques however allow spectral analysis to be undertaken.

MODEL APPLICATION

39. In the area of the study the movement of material plays an important part in the coastal management to be undertaken by the Consortium members. By a combination of visual observation, historical records, monitoring techniques and numerical modelling, a good understanding has been developed of the way in which the entity of the bay responds to natural and manmade influences.

40. The results of the numerical wave and tide model will determine the energy flux pattern at key locations for tide only, annual and extreme storm conditions. This, together with field measurements, will be used to define the general sediment movement, and hence the measures required to

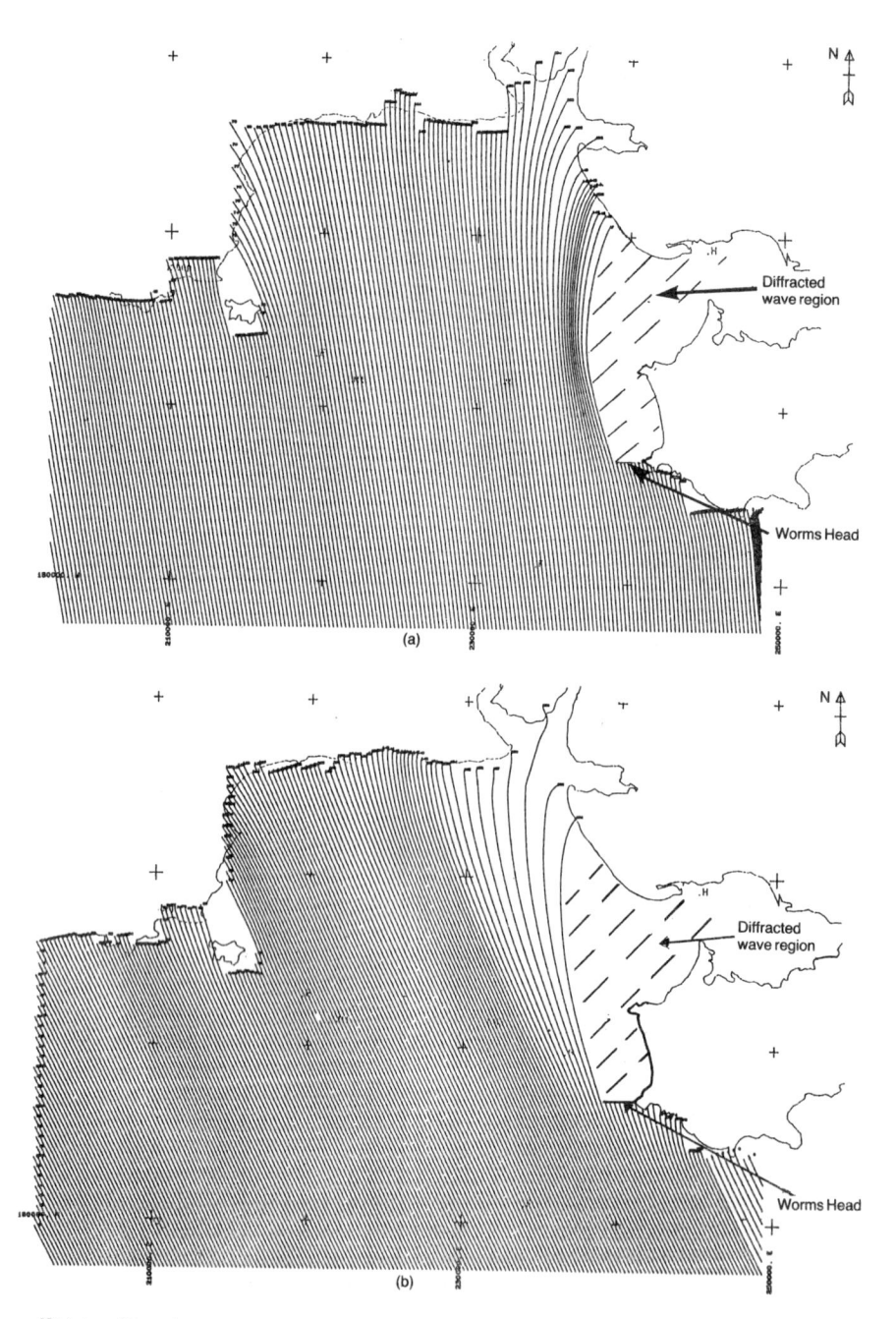

Note: Input ray angles are altered by the effect of current velocities offshore.

Fig. 6. (a) Wave Ray Plot, SSE Storm, M1 area (model A), high water -2 hours, showing effect of Worms Head, Carmarthen Bay Waveray Plot W1; (b) Wave Ray Plot, SSE Storm, Northern Estuary (M1 area), high water +2 hours, Carmarthen Bay Waveray Plot W1A

stabilise shoreline movement. Design parameters for necessary coastal structures (for example, the distribution of energy with direction) have also been determined for individual client's needs.

The present program of model runs undertaken is given below.

Bathymetry.   1977   1988 (part revision)

For each of the above, two tidal ranges have been examined:

   a)   Mean Spring Tide (MST)

   b)   Storm surge tide (based on storm 11th November 1977) (SST)

Wave conditions examined are:

   a)   South West (based on storm 11th November 1977) for SST

   b)   South West for MST (1988 bathymetry)

        South South West for MST (1988 bathymetry)

        South South East   "   "   "   "

The typical wave conditions offshore were obtained from the aforementioned wind wave model supplemented with observed swell conditions from the Meteorological office.

The results of the wave model show the following:

41. Changes in bathymetry between 1977 and 1988 have significantly changed the direction and energy attack at locations within the Burry Inlet and Northern Estuaries.

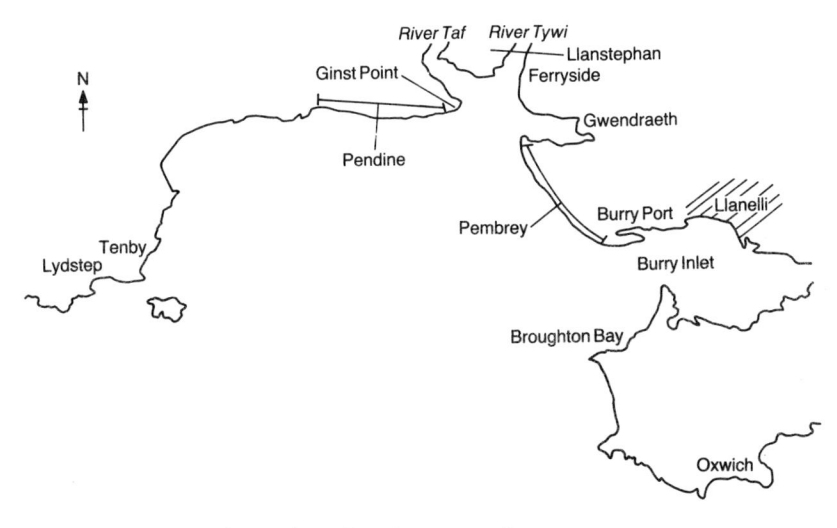

Fig. 7. Location plan for key results

ENGINEERING STUDIES

42. Inshore wave conditions are critically affected by not only the offshore wave direction (due to sheltering and local refraction effects) but also due to the magnitude and direction of tidal currents at the time of interest. The latter is illustrated by the wave plots for SSE storm condition (Hs = 2.7m Ts = 6.3 sec). For two stages of the tide (high water - 2 hrs and high water + 2 hrs shown in figure 6a and 6b respectively).

Through the use of such results long term trends may also be evaluated allowing a confidence in carrying out significant investment in the coastal area.

USE OF OUTPUT DATA

43. The Principal areas of concern and aspiration of the various involved Authorities are shown in figure 2. The study has permitted the competent examination of these interests and provides the Authorities with specific design guidance, where appropriate, as part of the final report. Transfer of study information comprises both report and software with the numerical model due for establishment on the computer at the Polytechnic of Wales. In this way the study results will be locally available for ongoing consultation and application. Individual projects requiring specific study need then to be referred back for specialist advice when they are under serious consideration. Monitoring work at a basic level is scheduled to continue so that study results may be further confirmed or modified as appropriate.

44. By such an approach to coastline management the clients have obtained a modern 'tool' to aid decision-making regarding development and conservation proposals affecting their specific shoreline and that of their neighbours within a recognisable 'coastal cell'. The numerical model study is therefore not discrete in time but forms an element in an ongoing process which is now organised to make more effective use of new data and understanding as it applies to the coastal zone in Carmarthen Bay.

KEY RESULTS

45. The study results have been extensive but in order to illustrate immediate benefits of this approach the following key results are presented (figure 7) -

    a) Pendine Dunes: Identification of key locations for dune stabilisation thereby avoiding high costs of 'linear' protection systems. Rationalisation of dune erosion to climatic conditions and long-term tidal cycle (19 yrs).

b) Ginst Point: Definition of environmental processes applying the influence of the river closure nearby. Establishment of alternative more cost-effective coastal works to stabilise the Point.

c) Ferryside/ Llanstephan: Determination of cyclic channel oscillation across the estuary creating specific exposure conditions at these sites overlaying a trend of estuarial change. Design criteria have been refined.

d) Taf/Tywi Estuary: Consequences of change due to closure of river Witchett and development at Ginst Point changing bank formation at Cefn Sidon and wave energy penetration up the estuaries. Trends are established and design criteria identified.

e) Pembrey Dunes: Identification of key locations for dune stabilisation thereby avoiding high costs of 'linear' protection systems. Rationalisation of dune erosion to climatic conditions and long-term tidal cycle (19 yrs).

f) Burry Inlet: Process understanding of regime changes and influence of training wall establishment and collapse. Clear identification of current trends and their influence upon design criteria for coastal works. Identification of potential strategic works to achieve several development objectives and economically permit developments not possible using 'local' solutions.

g) Broughton Bay: Understanding of reasons for unusual dune development and the exploration of recent trends - giving confidence in the efficiency of certain types of coastal works for application in the bay.

46. In addition to the site-specific results set out on the previous page it has been possible to establish key monitoring tasks the results of which will support, quantify or modify study conclusions to maintain future levels of awareness for consortium coastline managers. The study has also benefited from allied research investment by British Maritime Technology which has produced a new concept of sand-defence for the solution of dune control problems identified in Carmarthen Bay.

ENGINEERING STUDIES

ACKNOWLEDGEMENTS

The authors are indebted to the Authorities forming the consortium promoting this project for their permission to produce this paper and to the Polytechnic of Wales for their sound assistance in its execution. The authors are also indebted to the Department of Civil Engineering at the University of Liverpool, who developed the model software, and to Messrs Guthrie and Hodder of BMT Ceemaid Ltd who helped to prepare the paper.

REFERENCES

1) Barber P.C. (1985), 'Shallow Water Waves', ICE Breakwaters Conference.

2) Bijker, E.W. (1967), 'The Increase of Bed Shear in a Current due to Wave Motion', Proc. 10th Conf. on Coastal Engineering, ASCE, 1, 746-765.

3) Bouws, E and Battjes, J.A. (1982), 'A Monte Carlo Approach to Computation of Refraction of Water Waves' Jnl Geop. Res., 87 (C8), pp 5718-5722.

4) Christofferson, J.B. and Jonsson, I.G. (1979), 'Wave Action Conservation in a Dissipative Moving Fluid', Inst. Hydrodynamic and Hydraulic Engineering, ISVA, Tech.Univ, Denmark, 49, 21-29.

5) Flather, R.A and Heaps, N.F. (1975), 'Tidal Computations for Morecambe Bay', Geophys. J.R. Astro. Soc., 42, 489-517.

6) Kraus, N.C. (1987), 'Application of Portable Traps for Obtaining Point Measurements of Sediment Transport Rates in the Surf Zone', Jnl of Coastal Research, Vol 3, No 2, pp 139-152.

7) Owen, A. (1980), 'Tidal Regime of the British Channel: a Numerical Modelling Approach', Geophys. J.R. Astro, Soc., 62, 59-75.

8) Yoo, D. (1986), 'Mathematical Modelling of Wave - Current Interacted Flow in Shallow Waters', Ph D. Thesis, University of Manchester.

9) Yoo, D. and Hedges, T.S. (1985), 'Project Ceemaid: Wave and Tide Modelling, Formby Point'. Rpt No. MCE/DEC/85, Dept. of Civil Engineering, University of Liverpool.

10) Yoo, D., O'Connor, B.A. and Hedges, T.S. (1988), 'Numerical Modelling of Waves in an Estuary', presented at the 6th Congress of Asian Pacific Division of IAHR, Kyoto, Japan, July 1988.

21. Evolution of the Bournemouth defences

R. E. L. LELLIOTT, DipTE, MICE, MIHT, MInstWM, Borough Engineer and Surveyor, Bournemouth Borough Council

SYNOPSIS. Bournemouth Borough Council commenced the development of coast protection measures in 1907 with the initial phase of the Undercliff Drive promenade and seawall east of Bournemouth Pier. By 1919 some 6.5 km of the coastline had received protection but, due to the intervention of hostilities in 1939-1945, it was not until 1975 that the remaining lengths of coastline with landward residential development were provided with seawalls and promenades. As the philosophy of coast protection evolved, so have the methods used by Bournemouth engineers changed to reflect the art of the science.

EARLY HISTORY
1. The first section of the Undercliff Drive promenade and seawall linking the new East Cliff Lift and Bournemouth Pier Approach area was opened in 1907 and by 1911, the Undercliff Drive up to Boscombe Pier had been completed, as well as sections of promenade and seawall at Alum Chine and immediately west of Bournemouth Pier itself.
2. Already the seeds were sown for later generations of councillors and officers to try and deal with problems so created, since the construction of seawalls and promenades caused a cessation of the natural supply of beach material that would have otherwise fallen and maintained beach levels throughout a year. Fairly quickly in front of the new seawalls, beach levels started to fall, so much so that the provision of groynes was necessary to control the now well established losses in beach materials. Construction of groynes commenced in 1917 and by 1918, pre-cast concrete groynes were being built - demonstrating the expanding use of the new wonder material - concrete. By 1939, 29 No. concrete groynes provided rudimentary beach erosion control, including the massive Hengistbury Head Long Groyne.
3. Tables 1 and 2 indicate the progress of construction of seawalls, promenades and groynes, together with shape profile of the face of the seawall, and the construction of groynes. Table 1 shows Progress of Construction of Seawalls and Promenades in the period 1907-1975, and Table 2 indicates Progress of Construction of Groynes in the period 1917-1957.

ENGINEERING STUDIES

TABLE NO. 1 - PROGRESS OF CONSTRUCTION OF SEAWALLS AND PROMENADES 1907-1975

Date	Location	Length m	Shape Profile
1907	Undercliff Drive (Bournemouth Pier to East Cliff Lift)	700	Seawall - slope 1 in 2
1909	West Promenade (Bournemouth Pier to West Cliff Lift)	400	Seawall - slope 1 in 1
1909	Alum Promenade (Alum Chine to Borough Boundary)	455	Seawall - slope 1 in 1
1911	Undercliff Drive (East Cliff Lift to Boscombe Pier)	1510	Seawall - slope 1 in 2
1927	Boscombe Promenade (Boscombe Pier to Manor Bay)	825	Seawall - slope 1 in 1
1927	Boscombe Promenade (Fishermans Walk area)	455	Seawall - slope 1 in 2
1929	West Promenade (West Cliff Lift to Durley Chine)	570	Seawall - slope 1 in 2 (inc 100 m vertical)
1930	West Promenade (Durley Chine to Alum Chine)	487	Seawall - slope 1 in 2
1932/3	Boscombe Promenade (Manor Bay to Fishermans Walk)	815	Seawall - slope 1 in 2
1933/5	Southbourne Promenade (Fishermans Walk to Gordons Slip)	460	Seawall - slope 1 in 2
1955/8	Southbourne Promenade (Gordons Slip to Shell House)	380	Seawall - slope 1 in 1
1955/8	Solent Promenade (Bedfords Beach to Solent Road)	400	Seawall - slope 1 in 1
1957	Alum Promenade (rebuilding of storm damaged section)	200	Seawall - slope 1 in 1 with throwback
1962	Southbourne Promenade (Shell House to Bedfords Beach)	393	Seawall - slope 1 in 1
1972/5	Southbourne Promenade (Bedfords Beach area)	410	Seawall - slope 1 in 1

NOTES:

(i) All seawalls are counterfort and slab reinforced concrete with sacrificial facing of Purbeck limestone dubbers.

(ii) 550 m length of 1911 section of Undercliff Drive provided with toe piling and new facing 1968/71.

(iii) Apart from emergency repairs and (ii) above, only 1962 and 1972/5 sections of Southbourne Promenade are provided with toe piling to prevent undermining by sea in times of extreme beach scour.

(iv) Part of 1972/5 section of Southbourne Promenade now provided with stepped in-situ mass concrete facing instead of Purbeck limestone dubbers.

TABLE NO. 2 - PROGRESS OF CONSTRUCTION OF GROYNES 1917-1957

Date	Location	No	Length m	Construction
1917	Undercliff Drive-East Cliff Lift to Boscombe Pier	4	30	Precast concrete sections with rail piles
1917	Undercliff Drive-Boscombe Pier	2	45	Precast concrete sections with rail pile
1927	Boscombe Promenade	2	25/30	Mass concrete
1927	Boscombe Promenade	3	40/50	Precast concrete section with rail piles
1929	Durley Promenade	1	45	Mass concrete/outfall
1930	Durley Promenade	2	40	Mass concrete
1932	Boscombe Promenade	5	30/40	Mass concrete
1932	Boscombe Promenade Portman Ravine	1	52	Reinforced concrete with diamond head
1935	Boscombe Promenade Fishermans Walk	2	50	Reinforced concrete with diamond head
1935	Southbourne Promenade, Fishermans Walk	5	50	Reinforced concrete with diamond head
1937/9	Hengistbury Head Long Groyne	1	200	Mass concrete
1950	West Promenade West Cliff Lift area	1	45	Reinforced concrete
1951	Durley Promenade	1	45	Reinforced concrete
1952	Southbourne Promenade Works Scheme No. 1	3	30	Steel piling(Larssen No. 10A section)
1955	Southbourne Promenade Works Scheme No. 2	3	50	Reinforced concrete
1955	Bedfords Beach	1	40	Reinforced concrete
1955	Solent Promenade Southbourne	5	45	Reinforced concrete
1957	Alum Promenade	5	40	Reinforced concrete

NOTES:

(i) Wooden jetty at Alum Chine containing major surface water outfall converted to mass concrete groyne 1915.

(ii) Both Bournemouth Pier and Boscombe Pier have performed as permeable groynes.

(iii) Extremely hard driving conditions for sinking of piles in the period 1917-1957 caused great problems for stability of groyne structures.

ENGINEERING STUDIES

PROBLEMS WITH FALLING BEACH LEVELS

4. By 1957 the effects of lowering beach levels as a result of littoral drift (mainly to the east) and lack of supply provided by constant cliff erosion was beginning to generate concern amongst the tourism industry - especially as the 5 miles of "golden beaches" was not always there to be enjoyed! In addition, with reducing beach levels the lower apron and toe of the older seawalls were beginning to suffer from attrition with resultant breaches in the seawall itself and losses to promenades and associated infrastructure. The lower part of most seawalls was vertical and this in itself did not engender the favourable conditions for beach building since scour continued with the effect of the high reflectivity of the toe.

5. As a result of discussions and investigations with the then Hydraulics Research Establishment at Wallingford, trials were conducted with the experimental seagoing sand scraper in 1958. This was probably the earliest foray into beach renourishment whereby man recharged existing beaches using material dredged from offshore. Unfortunately, the prototype full-sized machine was not developed and Bournemouth still had a diminishing beach problem.

6. Monitoring and research into the criteria affecting the Bournemouth coastline commenced in 1951 and investigations were carried out into changes in offshore bathymetry by analysing Admiralty charts for the period 1911-1960, examination of Ordnance Survey sheets for the period 1870-1960, experiments with radioactive tracer elements to determine sediment movements, and recording beach levels at 53 points along the coastline between the Borough Boundary with Poole Borough Council and Solent Road, Southbourne. In the 11 year period 1951-1961, the highest average sand level was recorded in 1959 and the lowest in 1955.

INVESTIGATIONS AND RESEARCH, 1960-1972

7. In the period 1960-1972, Bournemouth Borough Council directly funded a considerable programme of research into coastal processes and offshore phenomena. The work was undertaken in consultation with the Hydraulics Research Station and embraced the following fields of study:-
   (a) seabed surveys - work by divers to determine the nature and extent of sediments, provide detailed contours and depth of sand in different areas.
   (b) seabed drifter and surface float experiments - confirmed the direction of drift, both at sea level and sea bed level, and to determine the extent of any recirculation of marine sediments within Poole Bay.
   (c) fluorescent tracer work - confirmed the presence and extent of the littoral currents which caused the drifting of sand, especially in the surf zone.
   (d) relationship between sea level and land levels in the Bournemouth area (before aerosols were heard of!) - investigations indicated that the rate of change will be

about 200-300 mm per century and has accounted for at least 7.6 m loss in beach width since the construction of the first seawalls.
(e) artificial seaweed trials - an area of seabed offshore at Southbourne was laid out with the first trial use of artificial seaweed utilising polypropylene fibres manufactured by ICI Fibers. Unfortunately, the trial was curtailed by D Jones Esq who organised a convenient storm to dislodge the 'unnatural stuff' from his locker.
(f) construction of experimental groynes - in conjunction with the Hydraulics Research Station 2 No. experimental precast concrete permeable groynes were constructed on the Undercliff Drive and have been proved to be highly successful, albeit rather expensive to construct. In addition, 7 No timber groynes of differing designs were constructed at Southbourne and from these was selected the present standard design of timber groyne.
(g) pilot beach renourishment scheme - an investigation was commissioned into the alternative methods of carrying out beach renourishment work, together with a close look into the most practical and economic sources of suitable marine sediments for this type of work. These investigations culminated into a trial scheme carried out in 1970 west of Bournemouth Pier, when some 84,500 m³ of dredged sand was placed largely about the low water mark of spring tides along a frontage of some 1800 m. The scheme was extensively monitored in conjunction with the Hydraulics Research Establishment and the experience gave confidence to the local coast protection engineers in order to prepare for a much greater scheme.

PREVAILING PROBLEMS

8. During the period of the extensive investigations and research the more pressing problems associated with ageing and crumbling seawalls still remained. Such was the extent of damages to the seawall on the Undercliff Drive between Bournemouth and Boscombe Piers, as a result of prevailing low beach levels, extensive reconstruction and strengthening of the walls, complete with toe piling was carried out over a length of 556 m in the period 1968-72.

9. In addition, erosion was continuing unabated at the eastern end of Poole Bay at a headland called Hengistbury Head. Here, the existing natural coastline was eroding at an average rate of 0.6-1.0 m per year with localised accelerations of 1.50 m per year. The beaches forming to the west of the Long Groyne were still growing with the western extremity of the extent of the new beach reaching to approximately Double Dykes, a length of 1200 m from the Long Groyne. Between this point and the existing seawall at Solent Road was a length of 1110 m of loose sand and gravel cliffs that was unprotected and suffering from continual erosion.

ENGINEERING STUDIES

TABLE No. 3 - GROYNE BUILDING PROGRAMME 1970-1990

Date	Stage	Location	No. of Groynes	Comments
1971/2	1	Southbourne Promenade (Shell House to Bedfords Beach)	3	Trial designs
1971/3	2	Southbourne Promenade (Shell House to Bedfords Beach)	2	Trial designs
1972/4	2A	Southbourne Promenade (Bedfords Beach area)	2	Trial designs
1974/5	3	West Promenade (Durley Chine area)	2	
1975	4	Undercliff Drive (West of Boscombe Pier)	2	
1977	5	Boscombe Promenade (East of Boscombe Pier)	2	
1977/8	6	West Promenade (Middle Chine to West Cliff Lift)	3	
1980/1	7,8	Undercliff Drive East of Bournemouth Pier	4	
1983	9	Boscombe Promenade (Honeycomb Chine area)	2	
1984/5	10	Boscombe Promenade (Manor Bay)	2	
1985/6	11,12	Boscombe Promenade (Portman Ravine)	4	
1986/7	13	Southbourne Promenade (East of Fishermans Walk)	5	
1987/8	14	Southbourne Promenade (West of Shell House)	3	
1988/9	15	Solent Promenade (East of Bedfords Beach)	5	
1989/90	16	Alum and West Promenades	5	

NOTES:
  (i)   Coast protection grant aid has been obtained for all Schemes up to Stage 15.
  (ii) Stage 16 not yet submitted to MAFF for approval - submission June 1989.
  (iii) Stage 13 used 5 No. differing hardwood planking timbers as a trial for MAFF.
  (iv) Stage 4 groynes have suffered terminal damage as a result of storm scour undermining.
  (v)   Stage 2A groyne to undergo extensive repairs as a result of 1987 hurricane damage.

More importantly, until 1985, the administration of the Coast Protection Act 1949 through the Department of the Environment prevented the issue of grant aid for coast protection works for what was then amenity land, with no element of landward development.

ADOPTION OF A STRATEGY FOR COAST PROTECTION

10. Armed with results of the investigations and research carried out in the period 1960-1972 and as a result of close consultation with the staff of the Hydraulics Research Establishment a strategy for coast protection was adopted by the Bournemouth Borough Council in September 1972. This policy embraced the abandonment of the proposed programme of seawall strengthening works and the initiation of the groyne building programme coupled with beach renourishment works. Further research work was to continue into the problems of providing cost protection works at Hengistbury head, east of Solent Road.

11. The groyne building programme commenced in May 1974 and has continued up and until the present time with the final phase due to be completed in financial year 1989/1990. The details of the programme of groyne building works are set out in Table 3. All timber groynes are constructed using greenhart timber piles and a variety of tropical hardwood timber planking. All fastenings and fixings are fabricated from non-ferrous materials in order to facilitate construction and reduce corrosion effects and to give a life expectancy of between 25-30 years.

12. The first major beach renourishment scheme was carried out in 1974-1975 and involved two types of operation:-
   (a) placing material offshore west of Bournemouth Pier over a length of 1800 m, at a distance of 200 m from the seawall. Some 106,000 m^3 of material was placed using this technique.
   (b) pumping ashore using a reclamation dredger the dumped cargoes of a primary trailer suction dredger over a length of 6500 m east of Bournemouth Pier up to the present end of the existing seawall at Southbourne. Approximately 658,000 m^3 of material was placed and measured on the beaches using this operation.

13. The works have been fully described by Willmington (ref. 1) and Newman (ref.2) both advocates of beach renourishment as a value for money method of tackling the diminishing beach syndrome and it is not intended to reiterate these authors' works in this paper.

14. Needless to say the 1974-1975 renourishment scheme was of sufficient national importance to promote Department of the Environment funded research into coastal processes in Poole and Christchurch Bays, as a natural extension of the work pioneered by Bournemouth Borough Council and adjoining coast protection authorities. This work involved the Hydraulic Research Station and Southampton University in the following fields of activity:-

ENGINEERING STUDIES

(a) Monitoring of Bournemouth beach renourishment scheme:
   (i) topographic and hydrographic surveying
   (ii) aerial photography
   (iii) project film
   (iv) volumetric analysis.
(b) Wave recording and analysis using waverider buoy sited in Poole Bay.
(c) Beach process research in Poole and Christchurch Bays.
(d) Development and performance of bastion schemes in Christchurch Bay.

15. The research project was managed by a Steering Group led by a representative of the Department of the Environment, Directorate General Water Engineering attended by representatives from Bournemouth, Christchurch and New Forest District Councils. The final report of the research project describes the various research studies that were co-ordinated by Sir William Halcrow and Partners (ref. 3). The project enabled the establishment of close liaison between the District Councils that has now continued through the aegis of the Standing Conference of Problems Associated with the Coastline (SCOPAC).

16. With the benefit of the monitoring established at the time of the Research Steering Group, further monitoring has continued on a six month basis such that total changes in beach volumes between mean high water mark of spring tides and 450 m offshore can be identified and are represented by Diagram No. 1. Volumetric analysis of the data provided under contract by Southampton University is carried out by Hydraulics Research Ltd on a fee basis using established main frame computer routines.

17. Based on the timetable for groyne building and monitoring of residual beach volumes the Council made provision for a third beach renourishment exercise in its capital programme for financial year 1991/92. However, the timing of these works changed radically in the summer of 1988 with the announcement by the Poole Harbour Commissioners of their intention to deepen the main navigation channel approach to Poole Harbour (see para. 36 et. seq.).

HENGISTBURY HEAD - A CASE FOR PROTECTION

18. As has already been mentioned, the steady erosion of Hengistbury Head had given concern to Bournemouth Borough Council councillors and officers for many years and numerous discussions with Department of Environment officials had only resulted in the construction of two timber permeable groynes at the end of the existing seawall in order to arrest the rate of erosion as the result of the classic terminal groyne situation. These groynes were constructed in 1976/77 and have resulted in a reasonable measure of control in the rate of cliff line attrition in their immediate vicinity - however, eastwards of their location the cliffs were eroding rapidly, so much so that there was concern about a possible breach of the cliff line east of Double Dykes and the creation of a

flood channel north eastwards into the southern side of Christchurch Harbour, thus making Hengistbury Head an island.

19. This potentially disastrous situation was brought to the attention of the Department of the Environment and with the strength of local feeling about the erosion of the Headland, not only on Solent Beach, but also north east of the Long Groyne, the Department of the Environment indicated that the strict policy on applying coast protection grant aid solely to developed, frontages and hinterlands was being allowed to relax.

20. Bournemouth Borough Council therefore took the opportunity to submit a package of coast protection measures early in 1985 to the Department of the Environment with a benefit-cost appraisal based upon procedures pioneered by Penning-Rowsell and Chatterton. However, in April 1985, the Ministry of Agriculture, Fisheries and Food assumed responsibility for the administration of the Coast Protection Act 1949 and so a new era in bureaucracy had begun.

21. Very quickly, the close working relationship already established with officials of the Department of the Environment in connection with coast protection matters was reforged with their colleagues in MAFF, especially the South Western District Regional Engineer's office.

22. The proposed package of measures included:-
(a) Model testing of proposed structures
(b) Construction of rock groynes on Solent Beach
(c) Construction of rock groynes north east of Long Groyne
(d) Reconstruction of causeway north east of Long Groyne to west of Mudeford Sandbank
(e) Renourishment of beaches on Solent Beach and north east of Long Groyne using shingle
(f) Construction of offshore breakwaters west and north east of Long Groyne
(g) Reconstruction of access road across headland upon completion of all works.

23. Upon Statutory Advertising, objections to various elements of the works proposals were received from Christchurch Borough Council and Mudeford and District Fishermen's Association. Christchurch Borough Council were initially concerned about the effects of the proposals for works north east of the Long Groyne upon the Mudeford Sandbank - especially as Christchurch Borough Council were responsible for coast protection works on the Sandbank. Mudeford and District Fishermen's Association objected to the proposals for siting offshore breakwaters west and north east of the Long Groyne and renourishment of beaches that utilised sea bed dumping techniques.

24. Negotiations with the Statutory Objectors permitted a withdrawal of objections in respect of model testing, groyne works on Solent Beach and reconstruction of the causeway north east of the Long Groyne. Because of the likelihood of a period of delay before the results of any model testing would be available, and by virtue of the urgency of the situation,

ENGINEERING STUDIES

works on Solent Beach were revised to consist of a Stage 1 that included:-

(a) Construction of two timber conventional groynes, east of the existing groyne field and

(b) Construction of emergency gabion and mattress protection to cliffs on either side of Double Dykes, at the lowest point in the cliff top profile.

25. Approval to the interim Stage 1 proposals was received in the autumn of 1985 and work commenced on Solent Beach in early 1986, with the model testing of the major structures undertaken by Hydraulics Research Ltd initiated in early autumn 1985. In the meanwhile, erosion of the coastline north east of the Long Groyne was well established with wholesale retreat of the coastline forming a deep embayment. More positive action was needed here, pending the outcome of the model testing at Hydraulics Research Ltd, and Bournemouth Borough Council responded by using its emergency powers under Section 5(6) of the Coast Protection Act 1949.

26. Work on executing the emergency works north east of the Long Groyne commenced in spring 1986 and involved the extension of the existing causeway across the established embayment utilising Portland stone block armour of 3-4 tonne weight on a mass concrete footing infilled behind with selected granular material. Smaller secondary armour of 1-3 tonne weight was placed against the primary stone to reduce the reflectivity of the new casuseway and the works were completed by early summer 1986. This left a length of unprotected cliff toe of 200 m up to the Long Groyne.

27. In the meantime, the Bournemouth Borough Council decided to seek the assistance of Middlesex Polytechnic Flood Hazard Research Unit to produce a full benefit cost appraisal that not only took into account tangible but also intangible benefits of conserving Hengistbury Head. The Headland occupies a unique position nationally by virtue of its archaeological and ecological importance, with unique features reflected by its Ancient Monument status and its designation as a Site of Special Scientific Interest.

28. By early summer 1986, the physical and mathematical modelling of alternative coast protection measures for Hengistbury Head had been completed and arrangements were made for the Statutory Objectors to visit Hydraulics Research Ltd to not only see how the proposed final solution performed but also to question Hydraulics Research Ltd scientists about the likely performance and side effects of the proposed works.

29. In the autumn of 1986, the final report of Hydraulics Research Ltd's investigations into the most beneficial solution to the erosion and protection problems at the Headland was accepted by Bournemouth Borough Council and forwarded to the Statutory Objectors. Further protracted negotiations were still required with the objectors, before each in turn withdrew its objection by March 1987.

30. However, the hurdles were not all cleared even at this stage. At the time of submitting the planning application for

the works a formal objection was received from the Nature Conservancy Council on geological grounds and this led to protracted negotiations about the management of the cliffs between the Long Groyne and the rest of the Mudeford Sandbank. These negotiations led to the Nature Conservancy Councils and the Bournemouth Borough Council entering into a Management Agreement under Section 15 of the Countryside Act 1968 for a length of the cliffs at the rear of the new causeway north east of the Long Groyne. This enable the Article 10 Direction or the Town Planning application implemented by the Department of the Environment to be withdrawn.

31. Approval to the final package of coast protection measures submitted to MAFF in 1986 was received in April 1987 and these approved proposals were:-
   (a) Completion of causeway north east of Long Groyne and construction of 5 No. rock groynes adjoining thereto:
   (b) provide gabion protection to ramp access and cliffs at the western end of Solent Beach
   (c) construction of 3 No. rock groynes at Double Dykes, extending existing groyne field eastwards
   (d) provide gabion protection to northern end of flood channel, south side of Christchurch Harbour, prevention of further erosion of the shoreline in this area
   (e) beach renourishment using shingle material on Solent Beach and north east of Long Groyne
   (f) stabilisation of cliffs above causeway north east of Long Groyne
   (g) carry out monitoring and coastal survey work.

32. Works on (a), (b), (c), (d) were commenced in late spring 1987 and finished in summer 1987. Some 24,000 tonnes of Portland and Purbeck stone were successfully hauled across the best part of Dorset without incident and through the conurbation of Poole and Bournemouth with a minimum of disturbance. With the completion of the 5 No. rock groynes north east of the Long Groyne, dramatic changes in beach levels were observed with the first groyne compartment north east of the Long Groyne saturated with beach material by natural means within 6 months of construction. Similarly, the other compartments in this groyne field all demonstrated rapid natural filling and despite the hurricane storm of 15/16 October 1987, beach levels have been maintained throughout the winter of 1987/88 and steadily improved through 1988.

33. Monitoring of the coastline regime using hydrographic and topographic surveying techniques has continued around Hengistbury Head, coupled with sediment sampling up to 400 m offshore. An extensive aerial survey of the Headland has also been carried out to establish the monitoring threshold in order to determine long term changes in beach and cliff profiles. Monitoring work will continue for a period of 5 years.

34. Beach renourishment of Solent Beach using coarse sized shingle commenced in the autumn of 1988 and was completed in late winter 1988. Some 143,000 m^3 of shingle was placed on

ENGINEERING STUDIES

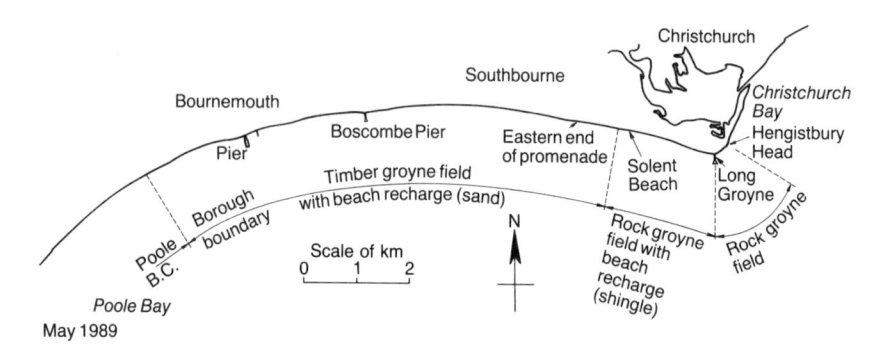

Fig. 1. Bournemouth Borough Council coastal management regimes 1989

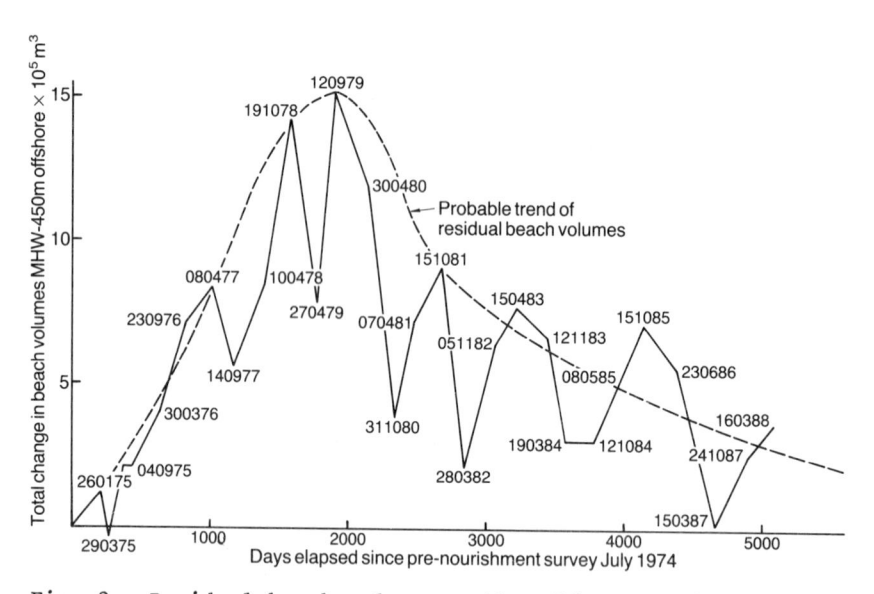

Fig. 2. Residual beach volumes - Alum Chine to Solent Beach

Solent Beach successfully, although the absence of groynes east of Double Dykes does give rise to concern about the long term stability of this section of the beach. Close liaison between Bournemouth Borough Council, MAFF Fisheries Inspectorate, local fishery interests and the contractor ensured an incident-free operation, despite the primary dredger being blown ashore and sustaining hull damage in October 1988.

35. The remaining element in the package of measures approved by MAFF will be the stabilisation of the cliffs above the causeway north east of the Long Groyne and this work will be carried out in autumn 1989, once the sand martins have left their habitual nesting sites in the upper part of the cliffs for their annual migration to Africa.

BEACH IMPROVEMENT SCHEME - STAGE 3

36. As mentioned in para. 17, early in summer 1988 the Poole Harbour Commissioners notified Bournemouth Borough Council of their intention of deepening the Swash Channel, the navigational approach to Poole Harbour entrance. Examination of sediment sample borehole logs and grading curves, together with details of the extent of the proposed dredging satisfied Bournemouth Borough Council that the dredged material was suitable for use as beach renourishment material.

37. Preliminary discussions with the MAFF Regional Engineer indicated that the proposed disposal of the dredging arisings by way of beach renourishment would provide a cost effective solution to two problems:-
  (a) A method of disposal of dredging arisings without affecting fishery interests.
  (b) Finding an economic source of suitable beach renourishment material that does not conflict with fishery interests.

38. Accordingly, Bournemouth Borough Council, having seen the economic advantage of carrying out a joint scheme with Poole Harbour Commissioners, resolved to press ahead with the next beach renourishment scheme - Beach Improvement Scheme Stage 3 - to be carried out at two locations:-
  (a) Borough boundary at Alum Chine, to West Cliff Zigzag, West Promenade.
  (b) Manor Bay, east of Boscombe Pier to Solent Road, Southbourne.

The estimated volume of material required to renourish these beaches was 632,000 m^3, derived from the regular monitoring beach profiles undertaken jointly between Bournemouth Borough Council and Southampton University.

39. Having received no formal objections as a result of statutory advertising and obtained all necessary consents, formal application was made to MAFF in August 1988, supported by a benefit cost appraisal produced by the Coastal Management Group of the Borough Engineer and Surveyor's Department of Bournemouth Borough Council. Formal approval to the scheme was received from MAFF in September 1988.

40. Successful negotiations with the Crown Estate Commissioners produced a satisfactory arrangement whereby the Crown Royalty levied on dredged material used for non-Estate benefit was fixed at an agreed level and only in respect of material placed on the beach.

41. Bournemouth Borough Council and Poole Harbour Commissioners negotiated and entered into a joint agreement in respect of their respective responsibilities since a common contractor was to be employed by both parties. These responsibilities are summarised as follows:-
   (a) Poole Harbour Commissioners - responsible for dredging Swash Channel and handing over dredger cargo at edge of Commissioners area of jurisdiction.
   (b) Bournemouth Borough Council - responsible for hauling dredger cargo to appropriate area, pumping ashore, distributing along coastline and grading to design profile.

42. Works commenced late in autumn 1988 and were scheduled to be completed by Easter 1989. Once again, close liaison was established between Bournemouth Borough Council, MAFF Fisheries Inspectorate, local fishery interests and the contractor to enable fishing activities to continue unabated, with the free passage of the primary dredger between the Swash Channel and the Bournemouth coastline.

FORWARD VIEW

43. Since 1907, all coast protection works in the Borough of Bournemouth have been designed by the staff of the Borough Engineer and Surveyor's Department under the instruction of myself or my predecessors (Lacey, Dolamore, Clowes, Whitaker, Vizard and Watson). The manner in which differing philosophies have been implemented over the years is a credit to the tenacity of these coast protection engineers and supporting staff serving the Authority. With the ever rising levels of public consultation necessary and bureaucracy from central government, who wants to be a coast protection engineer? These really are small challenges when compared with the resolution of the problems of managing the coast environment. We have not yet had an accurate assessment of the extent of the "Greenhouse effect" and the forecast rise in sea levels over the next 50-100 years.

44. For the immediate future, with the completion of the Groyne Building Programme in 1989 and the Beach Improvement Scheme Stage 3, the coast protection philosophy in Bournemouth would be virtually implemented. There may be a requirement to assess beach levels between Bournemouth Pier and Manor Bay, east of Boscombe Pier, and to examine the performance of the shingle beach on Solent Beach east of Double Dykes on Hengistbury Head. Thereafter, if sea levels are seen to be rising at more than the present predicted rate of 300 mm per century, then seawalls and promenades will need to be raised locally (and nationally) to tackle the overtopping problems so generated.

REFERENCES
1. WILLMINGTON R.H. The renourishment of Bournemouth beaches 1974-1975 Shoreline Protection - conference proceedings 14-15 September 1982. Thomas Telford Ltd, London, 1983.
2. NEWMAN D.E. A study in coast protection, report No. IT 174. Hydraulics Research Station, Wallingford, 1978.
3. - Poole and Christchurch Bays research project phase one report - volumes 1 and 2. Sir William Halcrow and Partners, Swindon, 1980.

22. Workington ironworks reclamation

B. EMPSALL, BSc, MICE, Projects Manager, Economic Development Department, Carlisle

SYNOPSIS. The British Steel Corporation foundry and ironworks at Workington closed in 1981 leaving 200 hectares of land and 4 km of coastline derelict. A major reclamation and redevelopment scheme has been carried out and this paper decribes the unique conditions on the coast, the alternative treatments that were considered, the sources of funding and the solutions adopted.

BACKGROUND
 1. In April 1981 the British Steel Corporation foundry and ironworks at Workington, on the west coast of Cumbria, closed leaving 1500 people redundant and 200 hectares of land, including 4 km of coast line, derelict. The effect on Workington was considerable and the County Council immediately resolved to find ways to repair the damage.
 2. Joint working parties were established with the Department of the Enviornment, the District Council and the British Steel Corporation, and following the preparation of a draft strategy and programme, the DoE agreed in principal to fund a substantial reclamation programme. The County Council itself also allocated £2m for works which would not be eligible for reclamation grant. The reclaimed area was to become known as Derwent Howe.
 3. The aims of the reclamation scheme were:
 (a) The improvement of the environment and image of Workington both as a town to live in and to make it more attractive to industrialists and
 (b) To bring the site into productive use for the creation of jobs and provision of recreational facilities for people and visitors.
 4. It was against this background that the treatment of the coastline was studied, debated and eventually carried out.

DESCRIPTION
 5. The coastline divides naturally into six sectors and these and other salient features are shown in figures 1 & 2.

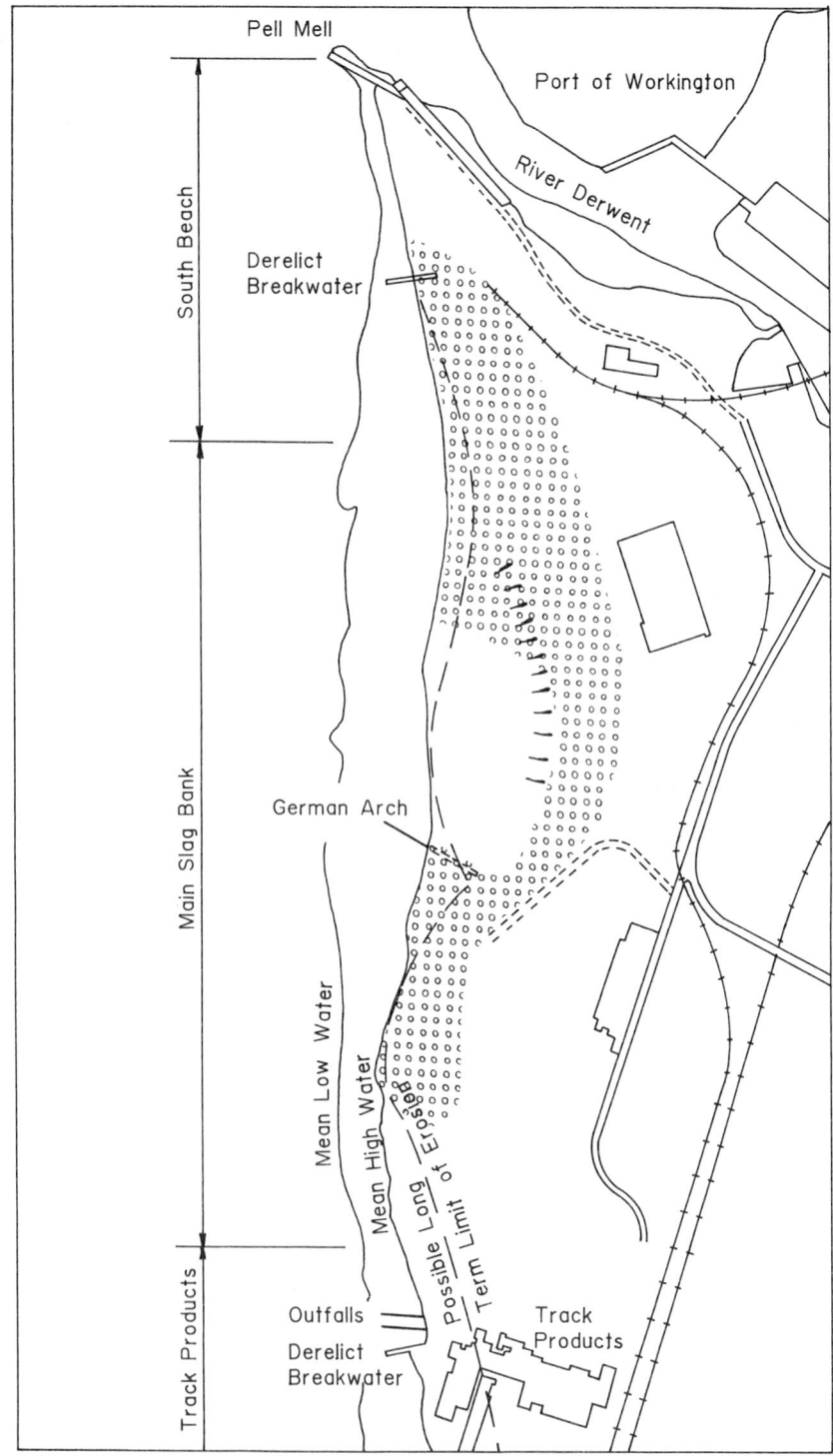

Fig. 1 Coastline - North

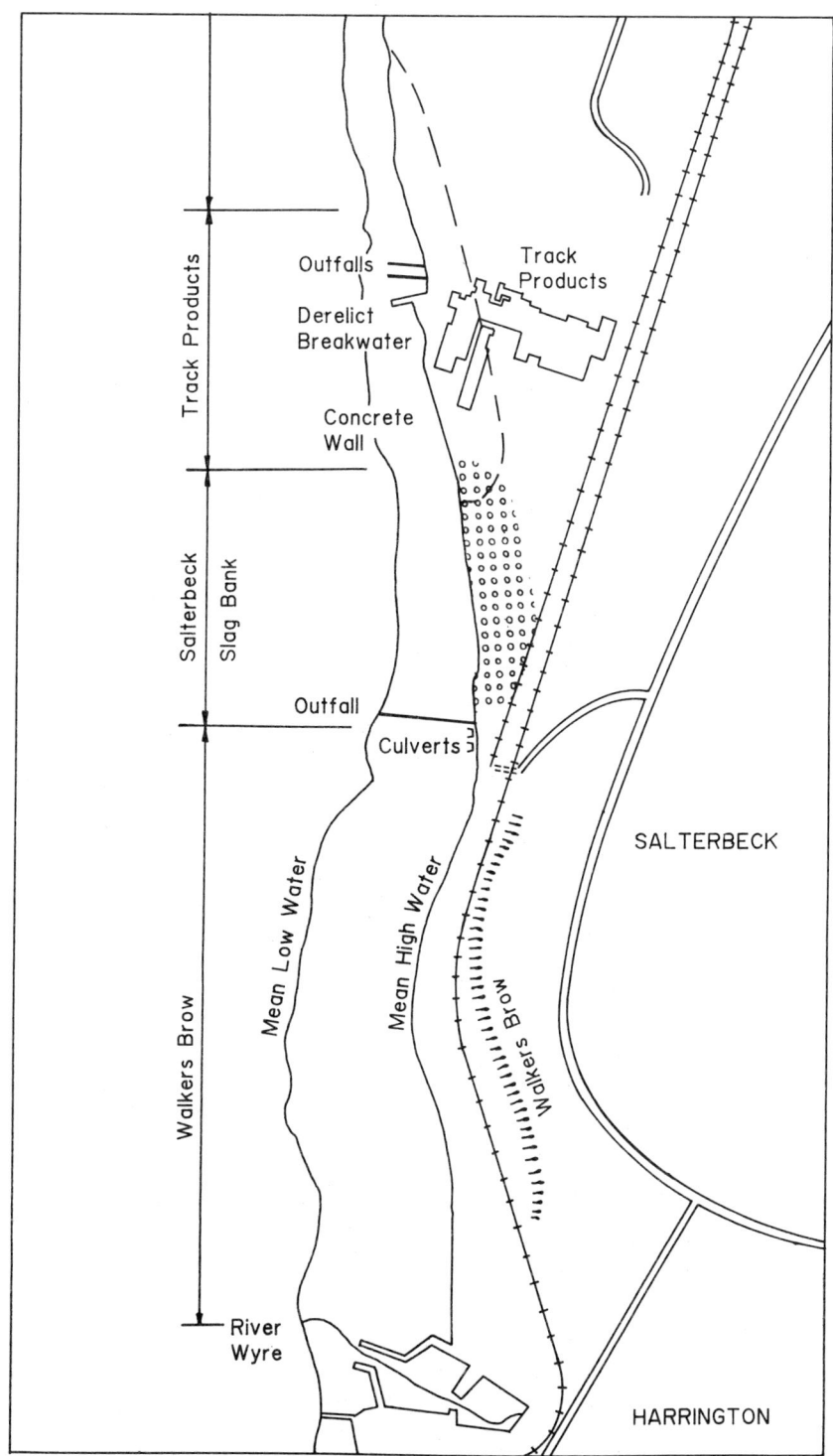

Fig. 2 Coastline - South

ENGINEERING STUDIES

Derwent River Frontage

6. This frontage extends from the Vanguard Sailing Club's tidal moorings to the Pell Mell breakwater and for the majority of its length comprises an attractive shingle beach with no erosion problem.

7. At the eastern end the beach is separated from the tidal moorings by a derelict stone pier remaining from the days when the area was a shipbuilding yard. Total collapse of this pier is likely to allow rapid siltation of the moorings thus destroying a valuable recreational asset.

8. At the west end of the beach the access road to the Coast Guard Station and South Beach is protected from erosion by 50 metres of derelict sea wall.

South Beach

9. South Beach is an easily accessible, attractive beach lying between Pell Mell breakwater and the beginning of the slag cliffs. Although at the north end of the site, it is known historically as South Beach because of its location immediately south of the river Derwent. This beach is divided in two by an old derelict breakwater.

10. The beach generally comprises mobile shingle but this overlies solidified slag which can be exposed by an adverse combination of wind and tides.

11. The large areas of solid slag which are a notable feature of this stretch of coast are formed by two distinct processes. In the first instance the waste slag from the iron and steel works was tipped, often molten, into the sea to eventually form a slag bank some 50 metres high. Where this has been eroded by the sea areas of solidified slag remain on the beach. The other areas of solid slag result from the cementing action which occurs when loose slag is immersed in sea water; this cemented material is commonly referred to as slagcrete.

12. Overall, material on this beach accretes rather than erodes but the sea was causing some localised erosion of the land behind the beach, especially south of the old breakwater. This localised erosion hampered access onto the beach and constantly exposed further dereliction.

13. This beach cannot be allowed to accrete indefinately as it will eventually outflank Pell Mell breakwater and add to the siltation problems of the Port of Workington.

Main Slag Bank

14. Originally the main feature of this sector was Chapel Hill (35 metres high) which had been eroded on its seaward side to form vegetated cliffs up to 30 metres high protected by a raised beach. As a result of iron and steel making Chapel Hill was buried under a slag bank which was 50 metres high and had been eroded to form sea cliffs up to 30 metres high.

15. The nature of the slag is not uniform and hence the rate of erosion of the cliffs is variable, ranging from a maximum of 3 metres annually by German Arch to practically

nil at headlands which are formed of almost pure iron. In
some places the sea is forming caves in the slag which are a
natural attraction particularly to children.

16. Except for isolated patches of sand the beach was much
despoiled and comprised solid slag and scrap metal.

17. The harder headlands bore some resemblance to natural
limestone cliffs and could be considered an asset to the
landscape. They support some wildlife, notably a pair of
peregrine falcons, and their retention was supported by the
Cumbria Trust for Nature Conservation and the Nature
Conservancy Council.

Track Products

18. At Track Products the factory frontage is protected
from the sea by a near vertical concrete wall terminating at
the south end in a steel sheet pile wall. To the north of
the wall protection was provided by the regular tipping of
concrete rubble which had been spread out over the beach by
the sea.

Salterbeck Slag Bank

19. This slag bank, which is much lower than the main
bank, has also been eroded to form cliffs, but here the
beach is more attractive being much sandier and the solid
slag less intrusive.

20. This slag bank affords Track Products with
considerable protection from the weather and BSC resisted
any proposal to lower it significantly.

Walkers Brow

21. Walkers Brow is the most sheltered section of the
frontage. The beach is variable, comprising areas of hard
slag, boulders, stone, shingle and sand. Erosion is minimal
and of no consequence except at the north end adjacent to
BSC land where protection has been provided by using blocks
of slag, skulls (the solidified rubbish which collects in
the bottom of ladles of molten metal) and other waste
material.

22. The beach is backed by the grassed embankment of a
disused railway line. Pedestrian access is available at the
north end and vehicular access from the south.

STUDIES AND INVESTIGATIONS

23. From the outset it was clear that before any sensible
options could be proposed for discussion, let alone firm
designs prepared, a substantial amount of basic information
needed to be assembled and Lewis and Duvivier were
commisioned to carry out a preliminary study. The object of
the study was to investigate the feasibility of possible
schemes of coastal protection work to prevent erosion and,
if practicable, to create amenity beaches.

Plan Information

24. Although no early specific surveys of the coast line
had been undertaken Ordnance Survey maps dated 1864,
1898/1900, 1924/25 and 1953/67 were available together with

one or two less reliable earlier plans. In addition the County Council had undertaken aerial surveys in 1982 and 1983 and these together with further aerial photographs dated 1948, 1958, 1973 and 1981 enabled an indication of the changes in the coastline to be established.

25. Examination of these records showed how the progressive dumping of slag and pier extensions had moved the shoreline seawards into deeper water by up to 200 metres and also changed the local alignment of the coast.

26. In an endeavour to establish some more upto date detailed information carefully surveyed metal pins were grouted into the foreshore at about 100 metre intervals and monitoring of the cliff toe relative to these undertaken.

27. From this and other information a prediction was made of the erosion which would occur in say the next 50 years if no preventative work was undertaken. This prediction is shown on figures 1 and 2.

Beach Levels

29. Understandably, historic level information was generally even scarcer than plan information except on South Beach. Some additional short term data was obtained by regularly checking the beach levels relative to the pins referred to in paragraph 26. Even this was not always straight forward as it was not uncommon to find that the pins, even though 25 mm diameter, had been laid flat on the beach by the action of waves and rocks.

30. Immediately north of South Beach lies the Port of Workington, now owned by the County Council, but formerly owned by the British Steel Corporation. Siltation of the channel into the port has long been a problem, with a significant bar forming north of the western end of Pell Mell breakwater. Regular expensive dredging is necessary to keep the port operational and BSC's predecessors the United Steel Corporation had undertaken a major study (ref 1), including the construction of a model, to identify training works which might be constructed to reduce the dredging requirements. As part of this study records had been kept of beach levels on South Beach over a number of years.

Beach Materials

31. Identification of the different areas of surface beach materials e.g. sand, mobile slag gravel, slagcrete, molten slag etc was carried out by onsite inspection and by careful study of the various aerial photographs.

32. As much of the existing beach material consisted of slag, and this was likely to remain the most readily available material for the foreseeable future, a series of slag attrition tests were undertaken. These were designed to assess the suitability of slag as a beach material by comparing the performance of slag under controlled conditions with that of other materials, such as quarry stone and beach shingle, as established in earlier experiments.

33. The tests were carried out by artificially abrading "as dug" slag until the maximum particle size was comparable to that typical of slag beach material at Workington and then further abrading the slag to establish whether attrition continued at the same rate.

34. Abrasion was carried out in a Stothert & Pitt 14/10 concrete mixer with all except two of the opposite mixing blades removed from inside the drum. All the samples under went three or four 3 hour tests and records were kept of description, weights and sieve analysis before and after the tests.

35. The results of the tests indicated that
   (a) the material would not survive long as a beach forming material,
   (b) there would be a tendency for the underlayers to cement into slagcrete thus destroying the permeable nature of a shingle beach it might be intended to provide and
   (c) the material would not appear to be mechanically harmful if placed on a foreshore, in that the end product is shingle, provided that it is placed in such a way as to minimise the rate of erosion of the placed material and therefore the rate of injection of fine material into the littoral system.

36. The stratas beneath the mobile beaches was determined by drilling boreholes. These revealed a slagcrete thickness of some 0.2 to 0.6 m overlying beach material.

Cliff Stability

37. The slag cliffs are the dominant feature of this length of coastline and their treatment was to be the subject of much debate. The three main issues were their aesthetic merit, safety of the public and their contribution to beach forming material.

38. The County Council's laboratory undertook a study to determine the stability of the cliffs. The slag heaps are far from uniform and comprise solid slag, reworked slag and miscellaneous tipped material. The density of the dry slag is approximately 1600 kg/m^3, roughly equivalent to some naturally occurring gravels. The cohesive resistance of the material is in the order of 50 kN/m^2 whilst the angle of internal friction is approximately 48°. Thus the slag displays the characteristics of both cohesive and granular soils.

39. The report concluded that the cliffs could not be retained as they were because even if protected from the sea they would still eventually fall back to their natural angle. They could not be preserved by conventional methods such as rock bolting or grouting. The recommendation was that the cliffs be regraded to an angle of 45° or less with the toe being protected from further erosion by the sea.

40. The County Planning Officer prepared a report arguing for the retention of the cliff on aesthetic grounds,

comparing them favourably with naturally occuring limestone cliffs and suggesting that if suitable ledges and crevices were created and vegetated the area could become a breeding ground for a diversity of birdlife.

41. The area was destined to become public open space and there was concern that members of the public could be injured in the event of a cliff collapse. the County Solicitor advised on the Council's liabilities.

42. The duty of the Council is to take such care as is reasonable in all the circumstances to see that the public do not suffer injury. The Council's duty to such persons does not necessarily involve the removal of the source of the danger: in deciding what is "reasonable" in the circumastances, the court would take account of the extreme high cost of doing so. The Council's duty to adults would probably be discharged by fencing off the cliff tops and erecting warning notices on the beach. However the safety of children would remain of concern. The courts have always expected an occupier of land to take greater care for the safety of children; childrent cannot be expected to read or abide by warnings and one must expect them to roam and explore. It is doubtful if the Council could ever discharge its duty to children (and thus remove the possibility of liability in the event of accident) without carrying works to remove the sources of danger.

Littoral Drift and Shoreline Geology

43. The BSC report referred to in paragraph 30 was the most significant source of information relating to sea bed conditions and this indicated that the sea bed material immediately offshore comprised a 'boulder scar' exposed over an area of the order of 1 km wide. On top of this is a narrow belt of mobile sand less than 400 metres wide immediately offshore and extending over the lower foreshore.

44. The information from the BSC report was augmented by a seabed survey undertaken in April 1983. Seventy six samples were taken from locations between 0 and 750 metres below low water mark in up to 10 metres of water. The locations were pre-located and the position determined on site by means of the Decca navigation system fitted to the boat. The samples were excavated by divers.

45. It was originally intended that both sieve analysis and chemical testing be carried out to assess the material's particle size and to identify any slag that may have been present. In the event however a preliminary examination of the samples indicated that very little, if any, slag was present in any of the samples and so only a grading analysis and visual examination was carried out.

46. The material fell predominately into two main categories;
   (a) cobbles and coarse gravel and
   (b) medium to fine sands. The uniformity co-efficients indicated all the materials to be single size with 96% of

samples having a uniformity co-efficient less than 5.

47. The coarser fractions of mobile beach material are moved primarily by wave action while the sand is moved by a combination of wave and current action. Material less than 0.1 mm diameter is generally carried offshore by currents to settle in deeper water. While the direction of beach material on the upper foreshore can reverse with oblique waves from different directions, this littoral drift is, on balance, in this area northwards under the dominant wave direction. The nearshore sea bed is dominated by the flood current so that on balance the movement of sand is also northward.

48. Much of the limited amount of coarse material moving into the area from the south is retained by groynes at Harrington and therefore the major source of this material along this frontage has to be from erosion of the slag bank. While some of the constituents of this are relatively hard others gradually wear away by attrition (see paragraphs 32-36). The general morthward movement of sand along the lower foreshore and nearshore zone is not restricted in the same way and it is reasonable to assume that a steady supply is available.

49. So far as the movement of beach material out of the study area is concerned, excavation of material from South Beach to prevent outflanking of Pell Mell breakwater (see paragraph 13) has effectively cut of any supply of coarse material to the north. It was estimated that this excavation amounted to less than 8,000 cubic metres per annum. The movement of the sand belt northwards is not restricted by the Pell Mell breakwater and it is this movement that leads to the siltation of the entracne to the Port of Workington. Of the order of 250,000 tonnes is dredged annually from the bar but this is likely to be only a fraction of the total quantity moving northward.

50. To complete the equation it would be necessary to know the rate at which the slag was deposited but unfortunately nor records were kept and in any case the reworking of the heap to recover the metal content and provide construction material would have reduced the value of any records.

FINANCING

51. As stated in paragraph 2 these coastal treatment works formed part of the Derwent Howe Reclamation Scheme which is being funded primarily through derelict land grant from the Department of the Environment.

52. There is no statutory definition of derelict land but for grant purposes the definition agreed with the Treasury is "Land so damaged by industrial or other development that it is incapable of beneficial use without treatment".

53. The grant is not available towards general development work and therefore any work on the coast had to be justified under the following headings:

ENGINEERING STUDIES

(a) Reclamation works: Works which are required primarily for the purpose of reclamation. As a general rule this includes all works necessary to bring the site to the equivalent of a green field site.
(b) New soil materials: The provision of new top soil where it is essential for the after use proposed eg public open space.
(c) Establishment of vegetation
(d) Basic Infrastructure: Cost of providing basic infrastructure which is ESSENTIAL for 'hard' development end use.
(e) Works necessary to protect the DoE's investment on the site.

54. In deciding whether or not to approve any particular scheme the Department considers local needs, value for money and the benefits and priorities of the scheme in relation to others. The need to satisfy these criteria was to have considerable influence on the works eventually carried out.

55. One other financial factor was to influence the design. Derelict land grant is a one off payment and is not available for the maintenance of works. It was therefore important that so far as possible the scheme should be maintenance free.

ALTERNATIVE SCHEMES

56. Before any detailed investigative work had been carried out there were four things which influenced thinking. First, the identification of the slag heap area as public open space led to requests from the planners for the creation of ammenity beaches. Second, the slag heap was known to contain 20 to 30 million tonnes of slag and any use of this which would help to reduce the height would be beneficial. Third, longshore drift was generally northwards and fourth, the creation of higher mobile beaches would help to protect the base of the cliffs.

57. Based on this information the first scheme tentatively proposed, in 1982, was for the formation of mobile beaches, both for ammentiy purposes and to provide soft protection, by reshaping the coastline to form bays which would restrict the natural movement of beach material. It was proposed that three equal sized bays should be formed between Track Products and Pell Mell breakwater. The bays were to be bounded by large promentories constructed of a landscaped slag infill whith appropriate formal protection. It was hoped that the northern bay in particular would provide an area of more sheltered water and the beach in the centre bay would protect the base of the cliffs.

58. Once the consultants had been appointed in 1983 and some initial investigative work had been undertaken it was decided to concentrate on three approaches to the problem. These were:
(a) Works necessary to protect the coastline and preserve

existing amenities and access.
(b) An intermediate solution that would provide enhanced protection together with a degree of improvement of amenity but having regard to cost in relation to benefit and
(c) Construction of major landscape structures to create bays as described in paragraph 57.

59. At this stage a discussion about the possible schemes was held with the planners and landscape architect and it was agreed that in principal:
(a) rock structures were likely to be visually preferable to more formal concrete structures and rock armoured rubble mounds preferable to rock and rail cribwork.
(b) The realignment of high water mark landwards should be avoided if at all possible.
(c) All schemes should include the clearance of scrap, demolition rubble etc. from the foreshore and
(d) If possible pedestrian access should be provided past Track Products. This was to form a link in the Cumbria Coastal Way.

60. Inevitably with so many interested parties it was not simply a matter of formulating three alternatives and choosing one; but rather a range of options was considered and an acceptable solution evolved. The following paragraphs outline some of the alternatives considered and the solutions eventually adopted.

Construction of Headlands

61. The construction of headlands was considered and broadly costed but although attractive on environmental grounds and in the amount of slag it would use the cost (over £8.5 m) ruled it out early on.

Derwent River Frontage

62. Erosion here is not a major problem although it is very desirable to repair the pier by the sailing club and tidy up the eastern end of the Pell Mell breakwater. Application for DLG for these works has yet to be made but it is likely that the majority of the funding will have to come from the County Council.

South Beach

63. The essential element of any work on South Beach was the need to prevent further erosion of the land at the back of the beach, as this would expose more and more dereliction. The stabilisation of these slopes could be achieved either by raising the beach levels or by constructing protection works in front of the slopes themselves.

64. The proposals for raising the beach levels involved the repairs of the existing breakwater, the construction of several substantial groynes and nourishment of the beach. Rock mound groynes were recommended as they provided good performance characteristics at reasonable cost and were acceptable aesthetic grounds. Timber or composite timber

and concrete groynes would have been simpler but to provide the necessary height they would have needed to be of heavy construction and therefore fairly expensive. They would also have had the disadvantage of being reflective under wave action and vulnerable to plank damage. Gabions would have been unsuitable since the wires are vulnerable to shingle abrasion and for the same reason sheet piling would have been rapidly perforated. Crib groynes were also considered but were aesthetically unacceptable and concrete structures would have been more expensive.

65. It was not felt that the beach would accrete sufficiently naturally and beach nourishment was proposed. Although an almost unlimited supply of slag was available locally tests (see paragraphs 33 - 36) proved it unsuitable and it was proposed to use the slag as hearting with an imported material on top.

66. The scheme eventually adopted was for the direct protection of the slopes themselves and the two alternatives proposed were either a simple rock armoured embankment or a stepped concrete wall topped by a low wave wall. The rock mound was the chosen solution as, being less reflective, it was more likely to encourage the collection and retention of mobile material, it was aesthetically acceptable and was less susceptible to foundation failure if any of the slagcrete was eroded by the sea. The land immediately behind the rock mound was turfed using turf reinforced with steel mesh. this has been found a successful method of establishing grass in an area which is subject to some wave action. This scheme was chosen as it was not only the cheapest but also the one most directly relevent to the reclamation works.

Main Slag Bank

67. The formation of any mobile beach in front of the majority of the main slag bank was not feasible because most of the existing beach was relatively smooth hard slagcrete and solidified slag. Some scope existed for the creation of beaches by the construction of groynes at German Arch and immediately north of Track Products. Even with these, some absorptive protection would be needed at the back of the beach.

68. Along the base of the cliffs, here and at Salterbeck, the protection proposed was either rock armouring or precast concrete blocks set and backed in concrete and keyed and/or dowelled in some way. Other alternatives considered were in situ concrete construction possibly in the form of a simple encasement of the toe of the cliff, or the application of sprayed concrete. Of particular concern here was the safety of construction workers in view of the unstable nature of the cliffs.

69. In the event finances and space dictated that the slag cliffs should in general be left untouched; with warning notices advising of the dangers being erected at strategic

locations. The only two areas receiving any treatment were those where the slag was particularly loose and forming unsightly scree runs. At these locations the slopes were regraded, soiled and seeded and the base of the slope stabilised by a rubble mound.

Track Products

70. Between Track Products and the main slag bank the dereliction (concrete rubble etc.) on the beach north of Track Products was removed and its coast defence qualities replaced by the construction of a rubble mound.

71. In front of the vertical walls a flat topped rubble mound was proposed to provide access and to reduce the reflective effect of the wall. It was felt that beaches could then be formed in this area by the repair of the existing breakwater and by the construction of further groynes. However except for the repair of the breakwater, which was paid for by the County Council, this work was not carried out as, being development rather than reclamation work, it was not eligible for grant.

Salterbeck Slag Bank

72. The proposals here were generally the same as on the main slag bank and for the same reasons little work was carried out except at the north end where it was necessary to support a haul road. Some attempt was also made to fill some of the voids here and on the main bank, by the use of cement filled bags and grout but the work proved difficult and ineffective.

Walkers Brow

73. There is no significant erosion occuring along this frontage and the only dereliction requiring treatment is at the north end adjacent to Track Products. Application will be made to the DoE for funds to pay for the replacment of derelict material with rubble mound protection.

DESIGN AND CONSTRUCTION

74. A conceptual design for the rubble mounds was provided by the consultants and the detailed design was carried out by the County Council. The design was conservative and conventional using the step by step design procedure laid down in the Shore Protection Manual (ref 2). The design prepared by the County Council was for a mound which relied entirely on the weight of stone for its stability. Following the recommendation of our consultants a bull head rail crib was added at the base of the slope to give added stability. The rock used on the surface ranged from 2.5 tonnes to 7.5 tonnes and the core was made of material ranging from 0.27 tonnes to 0.45 tonnes. The angle of the protection is 1 in 3.

75. Construction was carried out as part of a large contract which included reshaping the slag bank. Unfortunately the design team were not responsible for supervising the construction but generally the works

progressed without major problem. In places difficulty was experienced in founding the supports for the rail crib and this was overcome by excavating a strip foundation and setting the verticals in mass concrete.

76. In general the protection has performed well to date with the only signs of damage being near Track Products where, due to a hard outcrop, the slope exceeds the design of 1 in 3 and on a small area where the rock size is below specification.

ACKNOWLEDGEMENTS

The author is grateful to Mr John Burnet, The Director of Economic Development, Cumbria County Council for his permission to publish this paper. Thanks are also due to the authors colleagues and to Lewis and Duvivier (now Posford Duvivier) for their assistance in the investigations, design and construction and in the preparation of this report.

REFERENCES

1. UNITED STEEL CORPORATION. The silting of Workington Dock. 16 volumes published between 1946 and 1959.
2. U.S.A DEPARTMENT OF ARMY CORPS OF ENGINEERS. Shore Protection Manual, 1977.

23. Wirral scheme

C. D. DAVIES, BSc, MICE, Principal Engineer, Borough Engineer's Department, Metropolitan Borough of Wirral

SYNOPSIS. For the purposes of this paper the "Wirral Scheme" is considered to be the breakwater schemes constructed at Leasowe Bay and Kings Parade on the North Wirral coast. At Leasowe Bay two surface piercing, rock armour breakwaters were constructed during 1981 and 1982. Five submerged breakwaters faced with reinforced concrete Reef Blocks and constructed during 1984 and 1985, form the Kings Parade scheme. Both schemes were designed to stabilize beach levels by encouraging the natural deposition of beach materials. The two main performance aspects described in this paper are beach response and structural integrity.

INTRODUCTION
1. By the end of the 1970's the whole of the North Wirral coastline had been protected by man-made structures. The various structure types are shown in Fig. 1.
2. The coastline is fronted by a wide expanse of drying sandy foreshore which is most extensive to the west at East Hoyle Bank (Fig. 1). At low water sand banks extend to the edge of the training banks to the Crosby Channel which is the only dredged channel serving the Port of Liverpool. Littoral drift is in a predominantly N.E. direction, the major supply of sediment to the North Wirral coast is believed to come from the North Wales coast.
3. East Hoyle Bank affords a degree of protection to the western half of the North Wirral coast. However, the remainder of this coastline is subjected to a combination of a high tidal range on spring tides and severe exposure to waves generated during NW gales. The principal physical parameters affecting the coastline are shown in Table 1.
4. Reconstruction of the existing coastal structures had commenced in 1972 and was due to continue for 15 years. By 1975, a detailed analysis of existing records had shown that the North Wirral foreshore was subject to a general erosion that had been developing over the previous twenty years. Beach erosion had also been recorded along the North Wales coast indicating a deficiency in the supply of sediment on to the North Wirral coast. It was realised that if beach levels were not stabilized the new and existing defences would deteriorate faster as beach levels dropped. Serious consideration was now given to estab-

ENGINEERING STUDIES

Fig. 1. Location Plan

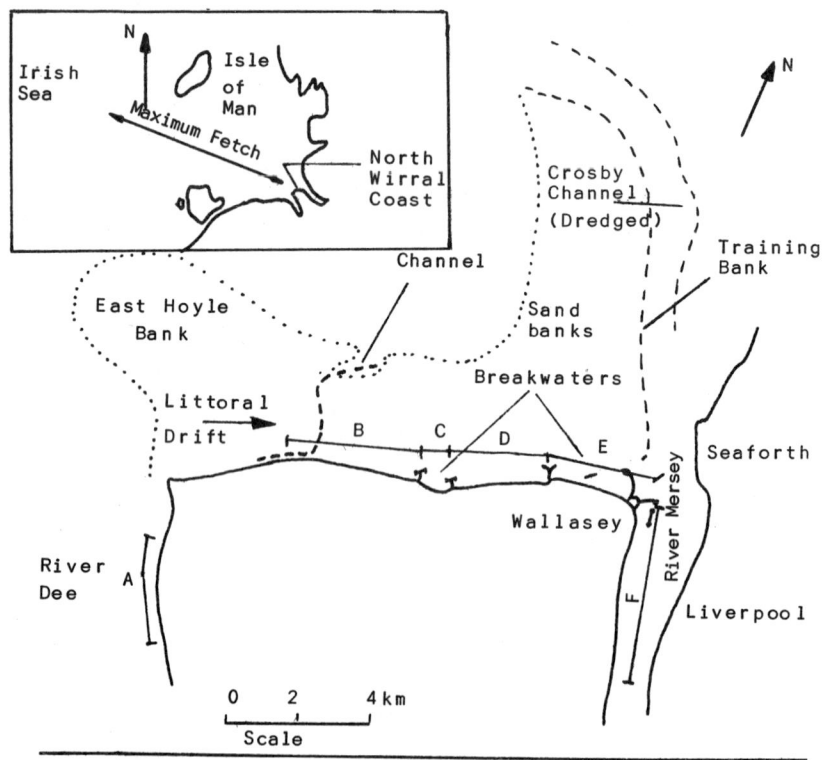

Location	Construction	Slope
A. River Dee coast	Rip-rap	1:2
B. Wallasey embankment	Concrete T-units	1:6 to 1:3.75
C. Leasowe Bay	Rip-rap	1:2
D. Leasowe revetment	Concrete T-units	1:3.75
E. King's Parade	Concrete	Vertical
F. Mersey river wall	Sandstone	Vertical

Table 1. Physical Parameters

Maximum Tidal Range	10.0m
Maximum recorded tide level	6.09m aod
Maximum fetch	180km from WNW
Annual maximum wave height at the Mersey Bar	10.0m
Inshore wave height*	2.5m ± 0.5m
Wave period	7-9s

*Typical observed heights in storm conditions

PAPER 23: DAVIES

lishing schemes which would stabilize beach levels and thus secure the existing and proposed coastal structures.

LEASOWE BAY

5. The historical, feasibility, administrative, design, construction, financial and monitoring aspects of the offshore breakwaters at Leasowe Bay have been presented in the Proceedings of the Institution of Civil Engineer's (ref. 1). For completeness, however, the following paragraphs refer briefly to some important aspects.

6. Leasowe Bay (Figs. 2 and 3) is not natural in occurrence but has developed as a direct consequence of the construction, either side, of Wallasey Embankment and Leasowe Revetment.

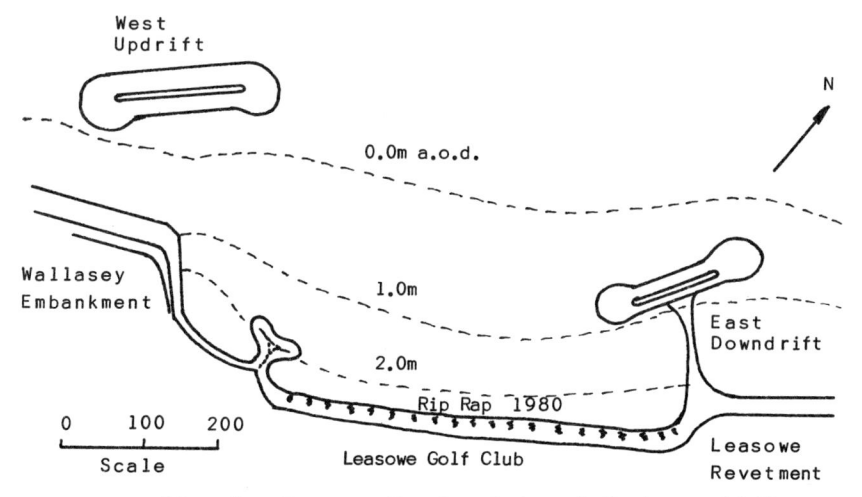

Fig. 2. Leasowe Bay Beach Level Contours 1980

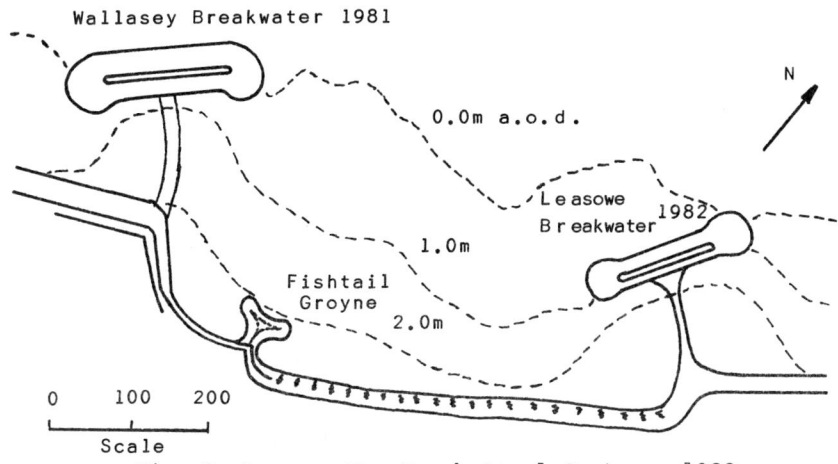

Fig. 3. Leasowe Bay Beach Level Contours 1988

295

ENGINEERING STUDIES

7. At the beginning of the 1970's it was felt that the problems of coastal defence in Leasowe Bay would best be solved by linking the two existing sea walls with a "hard" structure. However by the end of the decade there was a greater understanding of the local processes causing erosion and damage. A solution was sought that would reduce wave energy within the bay area including its extremities, by accentuating the bay affect.

8. These performance requirements were satisfied by providing offshore breakwaters at either end of the bay (Figs. 2 and 3). The main breakwaters were constructed in rock armour imported from limestone quarries in North Wales. The predicted reduction in energy levels within the bay allowed the inner bay to be protected by a 1 in 2 rock armour revetment. A small fishtail groyne was also provided at the end of the concrete return wall to Wallasey Embankment.

9. The original proposals for the updrift breakwater included a shore-connected link. However, concern was expressed with regard to its effects on the littoral drift to the east. As the shore-connected link could always be introduced at a later stage it was decided to initially provide a detached breakwater. The shore-connected link to this breakwater was finally constructed at the beginning of 1987.

KINGS PARADE BEACH STABILIZATION (refs. 3-5).

10. To date no detailed technical papers have been prepared on the Kings Parade scheme although it was referred to in outline. in ref. 6. It is proposed to prepare such a paper towards the end of 1989 when several years performance data will be available

11. Kings Parade sea wall is a 3.5km long mass concrete vertical gravity wall which is founded on either the sandstone or boulder clay which underlies the sandy foreshore in the area. The wall was built as an Employment Relief Scheme during the 1930's on a line up to 300m seaward of the natural coastline. It was soon evident, that the increased energy levels caused by the position and shape of the new structure were causing serious beach erosion. Beach levels taken regularly since the time of construction indicated by 1970 that levels in front of the wall had fallen by up to 4.0m. If this erosion had been allowed to continue the foundations of the wall would, within a few years, have been at risk from undermining.

12. In 1975 a research programme was instigated into the causes of beach erosion at Kings Parade. Firstly a data base was established and used to produce a numerical model of water motion. This model predicted sediment transport by wave and tide forces. The model predictions were then verified and improved by the use of specifically collected field data. A wave rider buoy established at the Mersey Bar provided details of offshore wave climate. Wave direction in the near shore zone was obtained from a land based radar installation. An instrumentation tower erected on the beach in front of the sea wall provided information on wave climate and current regime.

13. Traditional methods of solving the erosion problem at Kings Parade had been examined, however these would have merely secured the foundations of the wall. The long term erosion problem would not have been arrested. By using the numerical model the most sensitive locations and effective forms for protective breakwaters were identified along Kings Parade (Figs. 4 and 5).

14. The proposed structures at Locations 1, 3 and 4 (shore connected) were designed to divert tidal flow offshore and to intercept the principal wave trains and their respective reflective components off the sea wall. The structures at Locations 2 and 4 (offshore) were designed to intercept wave activity, with the latter structure intercepting waves reflected off the dock walls at Seaforth.

15. The required performance characteristics of the proposed structures were established from the research and observation of the performance of the Leasowe Bay breakwaters. It was determined that the structure would be submerged and its form

Fig. 4. Kings Parade Beach Level Contours 1979

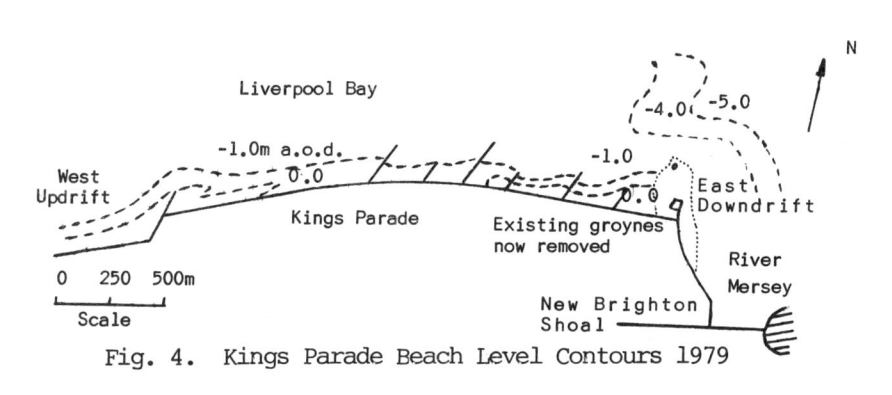

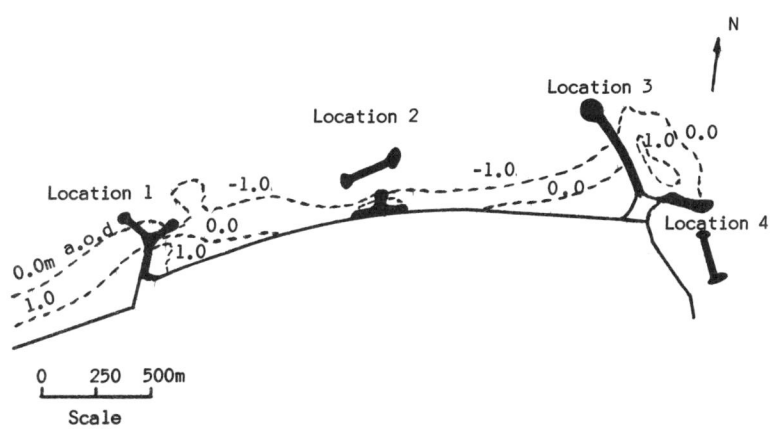

Fig. 5. Kings Parade Beach Level Contours 1988

ENGINEERING STUDIES

would allow flow through the upper 2.0m. This would help to prevent the seaward set up of stillwater level and ensure that regime changes took place gradually. It would also lead to cost savings in the greater water depths at Kings Parade.

16. These performance requirements would have been difficult to obtain with an armour stone structure, due to its limited porosity, stability and susceptibility to damage on the back face during overtopping. The structure of the breakwater utilises both armourstone and precast concrete "Reef Blocks" and was developed from physical model studies at both Liverpool University (to provide guidelines for block shape), and HRL Wallingford (to determine the crest heights and widths of the structures).

17. The use of a "hybrid" construction, in concrete and armourstone, allowed the stable construction of steeper gradients and a lower crest level. This in turn considerably reduced the volume of material per metre run of breakwater, making a hybrid solution cost effective in comparison to a rock armour structure. The final cost of the scheme was also over 20% below the estimated cost of merely protecting the toe of the wall.

18. Breakwater construction commenced at Location 2 during the spring of 1984. By the end of that year the structures at Location 3 and 4 had been completed. The final updrift structure at Location 1 was completed in spring 1985.

19. The purpose of the Kings Parade breakwaters was, by their effect on the wave and current regime, to encourage the deposition of beach material within the envelope formed by the breakwaters. This deposition would lead to a reversal of the long term erosion, protect the foundations of the wall and improve the amenity value of the beach. An increase in beach levels at Kings Parade would also encourage accretion on updrift sections of coastline.

MONITORING

20. It is, important to monitor the effects of any new coastal structure. The two main areas of concern are normally the effect the structure has on the coastal regime and the structural performance of the fabric of the structure. Such monitoring is essential on breakwater structures especially those which are designed to influence beach levels by affecting the natural regime. It is equally important to be patient in reacting to the results of the early monitoring of such structures. Their influence is likely to establish gradually over a period of several years.

21. The monitoring system developed as relevant to defining and analysing the effect of the breakwater schemes consists of the following:-

(a) annual beach level measurement developed to provide additional information in the area of the breakwaters
(b) investigation of bed levels seaward of the breakwaters by taking soundings
(c) the provision of an on-shore and off-shore wind recording system to allow current and future conditions

to be compared against a base model established from previous records (ref. 7)
(d) float studies in the areas of influences of the breakwaters
(e) use of aerial photographs taken biennially and the development of the use of obliques taken on a more regular basis to identify major changes in beach and channel configuration
(f) observation of outside influences (e.g. dredging activities) on changes in natural features particularly in updrift areas
(g) regular inspection of rock armour and concrete unit structures, augmented by durability monitoring using electrical methods of a sample of concrete units. Co-ordinated and levelled reference points also established on structures to allow more detailed checking.
(h) consideration of the environmental, ecological, amenity, leisure and safety aspects of the structures and reaction to any problems

LEASOWE BAY MONITORING

22. Reference 1 contains a section on the monitoring of the updrift breakwater at Leasowe Bay in the eighteen months immediately after construction. This monitoring was supported by SERC and undertaken by the Authority in conjunction with Liverpool University. Delegates are referred to paragraphs 30, 31 and 32 (ref. 1) for details of the conclusions resulting from that monitoring work. The monitoring and performance information given in this paper complements that given in the above reference.

23. Records of coastline recession and beach levels in Leasowe Bay date back 100 years. Beach levels having been taken regularly over the areas of influence of the Leasowe Bay Breakwaters for over 20 years. These levels are taken along fixed sections extending up to 500m from the coastline. The sections are approximately 150m apart. Since breakwater construction the sections have been augmented by a grid of levels which covers the areas around the breakwaters and the bay area.

24. The level information has been analysed in terms of areas above specific contours and bulk volumes above a datum. It is considered that in this instance, area change is a better indicator of breakwater behaviour than bulk volume. Volumetric analysis may mask the fact that a great deal of recorded accretion has in fact taken place in the breakwater lee within the resultant tombolo. At Wirral, measurement systems are being developed which use vertical and oblique aerial photographs which will complement the area analysis method. Finally, in comparison to volumetric analysis, the area method is simple, quick and requires fewer approximations.

25. Only the overall changes in beach levels in the area of Leasowe Bay, and the immediate updrift and downdrift sections are presented in this paper. Figs. 2 and 3 show pre-construction and current position of the 0.0, 1.0 and 2.0m contours.

ENGINEERING STUDIES

26. If the area of influence of the breakwaters is considered as three sections, namely updrift of the west breakwater the bay area (between the breakwater roots) and downdrift of the east breakwater, an understanding of the effect of the breakwaters on the beach can be obtained.

Updrift section - Leasowe Bay

27. Since construction the total area above the 0.0m contour has more than doubled along the 650m length of Wallasey Embankment. An area above the 1.0m has now been established against the toe of the embankment. This updrift area of accretion is now playing an important role in helping to control the scouring effect of the ebb channel which flows from behind East Hoyle Bank (Fig. 1). As a result of an eastwards movement of East Hoyle Bank in recent years this channel has moved landward and eastwards. It runs along the toe of the western section of Wallasey Embankment. One of the main reasons for adding the shore connected link (top level 2.5m) to the western breakwater was to help maintain higher beach levels immediately updrift.

Bay area - Leasowe Bay

28. Between the two breakwaters there has been a significant increase in sand quantities as indicated in Table 2. The main changes in contours being that the 1.0m contour now runs from the western breakwater whereas it was previously controlled by the end of the embankment, where the 2.0m contour has now been established. Beach levels in the lee of the western breakwater have increased by more than 1.0m. Accretion has been particularly impressive in front of the concrete return wall to Wallasey Embankment.

Table 2. Area Changes - Leasowe Bay (hectares)

Area above	1980	1988	Change	%
a) Area to 650m west of updrift breakwater				
0.0m aod	4.33	9.20	4.87	+112.0
1.0m	-	2.37	2.37	-
b) Total areas between breakwater link arms				
0.0m, aod	18.79	21.55	+2.76	+14.7
1.0m	7.5	10.09	+2.59	+34.5
2.0m	2.76	5.55	+2.79	+101.00
c) Area to 800m east of downdrift breakwater				
0.0m aod	14.85	13.15	-1.70	+11.50
0.5m	9.85	8.25	-1.60	-16.25
1.0m	7.80	3.80	-4.00	-51.25
2.0m	0.44	0.75	+0.31	+70.45
Area above	1980	1982	Change	%
0.5m	9.85	7.80	-2.05	-20.80
1.0m	7.80	3.40	-4.40	-56.40
Area above	1982	1988	Change	%
0.5m	7.80	8.28	+0.48	+ 6.15
1.0m	3.40	3.80	+0.40	+11.80

29. Within the bay area the beach has generally become flatter as the contours have moved seaward. The beach level has only risen marginally against the rock armour wall. However previously the beach against the coastline was kept at an artificially high level by the presence of clay and brick washed out from the earlier attempts to protect the coastline. The small increase in beach levels in this area has been sufficient to cover the brick and clay.

30. In the lee of the eastern breakwater beach levels have risen by over 2.0m. Beach accretion has taken place almost symmetrically about the submerged link constructed at a top level of 3.0m. At the western breakwater the tombolo shape and position, in the absence of a shore connected link initially, has been modified by a combination of tidal and wave induced currents. The formation of the tombolo to the eastern breakwater took place over a period of two years, without the deposition of large areas of silt which had been a feature of the early effects of the first breakwater. Sand now covers the root of the link at all times, and extends over most of the link during the summer. Initially the summit of the tombolo coincided with the link area, but as the area has continued to fill with sand the summit of the tombolo has started to move downdrift. Beach development has now reached the stage where the natural energy regime is the major influence. Previously the shore connected link had artificially controlled beach development in the area. These recent developments should ensure an increased level of material supply downdrift from the lee of the breakwater.

Downdrift section - Leasowe Bay

31. Table 2 provides information on the performance of the downdrift section of the beach since construction of the breakwaters. Inspection of the area changes would indicate that breakwater construction has dramatically starved the downdrift beach. It had been predicted that a solid link to the breakwater might divert the littoral drift offshore for over 600m downdrift of the structure. It was therefore decided to introduce a submerged porous link. The contractor elected to gain access to the breakwater by temporary elevation of the shore-connected link using stone fill. For six months during summer 1982 the link construction was similar to that which had been predicted to divert the littoral drift offshore. These predictions were confirmed as the downdrift beach level dropped by up to 1.0m. Table 2 shows that the final breakwater construction, with the submerged porous link, has not led to starvation downdrift. In fact sand is starting to accrete in the downdrift section, coinciding with the movement of the tombolo summit referred to in pargraph 30.

Conclusion - Leasowe Bay

32. The Leasowe Bay Breakwaters have performed well in increasing beach levels within the bay area and securing both ends of the bay. Energy levels have reduced to such a degree

ENGINEERING STUDIES

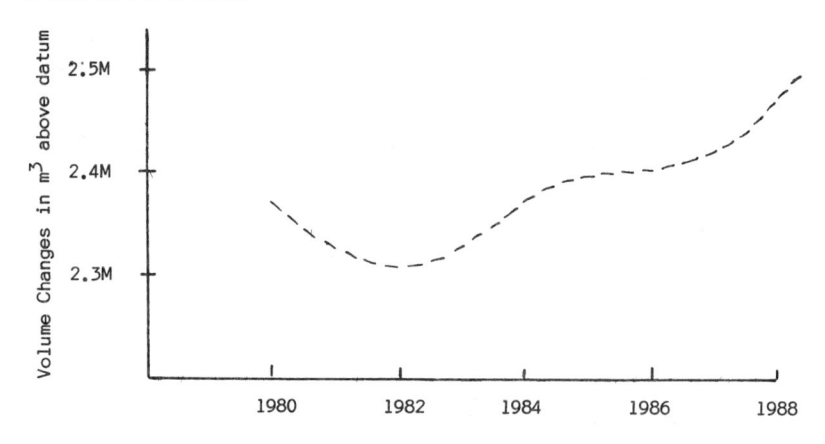

Fig. 6. Leasowe Bay Volumetric Changes
(over 60 hectares)

at the inshore wall at Wallasey Embankment and in the lee of the downdrift breakwater, that high beach levels now exist which are ️dry on MHWN tides. The damaging edge waves and plunging waves at the return to Wallasey Embankment have been eliminated. At the eastern end of the bay, the end of Leasowe Revetment has now been secured by the newly established high beach and rock armour protection.

33. Although the downdrift beach levels are still recovering from the effects of the temporary diversion of the littoral drift, the foundations of the newly constructed revetment are not at risk. The beach profile on this downdrift section has improved, the deep ebb channel which used to flow from Leasowe Bay past the corner of Leasowe Revetment has now been eliminated.

34. The figures presented in Table 2 give comparisons for two years 1980 and 1988. As such this is very much a "snap shot" view. However when considered with personal observations and photographic records, the information fairly represents current trends. Beach levels in the area had fallen continuously during the 1970's especially towards the end of the decade and in fact the 1980 levels were a slight peak on a downward trend. Figure 6 indicates the volumetric changes in an area extending to 500m seaward of the coastline, and 500m to either side of Leasowe Bay. Although wind and storm records for the period since construction have not been analysed there is a local view that energy input from the N. to W. quadrant during the 1980's has to date been at a reduced level when compared with the 1970's. This has probably resulted in a lower rate of supply of beach material into the area of influence of the breakwaters.

KINGS PARADE MONITORING

35. The Kings Parade Beach Stabilization scheme has to date only been exposed to four summers and three winters. Therefore

a detailed analysis of beach response has not been completed as yet. Accordingly a brief assessment of the beach level trends has been prepared for this paper.

36. Fig. 4 shows the preconstruction contours for the Kings Parade frontage, these contours were artificially controlled by a series of groynes constructed in 1953. The results of the most recent beach level survey in 1988 are shown on Fig. 5. Beach levels in the area of Kings Parade have been recorded at fixed cross sections since the construction of the original wall in the 1930's. The original sections have now been supplemented by a grid of levels over the whole frontage. All levels are recorded annually during the summer.

37. It was predicted that beach development at Kings Parade would take place initially at the updrift section around Location 1 and also in the area of Location 3. It was also predicted that the combined effects of Location 1 and Location 2 would initially cause starvation immediately downdrift of Location 2.

38. The areas above the -1.0, 0.0 and 1.0m contour have been calculated for the length of influence of the Kings Parade breakwaters excluding Location 4 offshore. A small scale beach nourishment operation, promoted by the Authority's Tourism section, has been carried out recently in the lee of location 4 offshore breakwater. This has made the determination of this breakwater's effect on naturally developed beach levels difficult to determine.

39. Considering the main Kings Parade frontage which lies between Location 1 and Location 3 it was predicted that new higher beach contours would be controlled by the two breakwaters. Location 2 provides a central control point to ensure that the final contours do not run into the existing wall. Inspection of the contour plans and the area information provided in Table 3 shows that the trend is in accordance with design predictions. The development immediately downdrift of Location 1 took place very rapidly as a result of a combination

Table 3. Area Changes - Kings Parade (hectares)

Area above	1979	1988	Change	%
Total areas to 500m west of Location 1				
-1.0m aod	16.70	18.90	+2.20	+13.20
0.0m	10.80	15.05	+4.25	+39.35
1.0m	4.80	8.10	+3.30	+68.75
Total areas between Location 1 and Location 2 breakwaters				
-1.0m aod	13.16	17.00	+3.84	+29.20
0.0m	4.67	3.16	-1.51	-32.35
1.0m	-	1.08	1.08	-
Total areas between Location 2 and Location 3 breakwaters				
-1.0m aod	16.08	14.95	-1.13	- 7
0.0m	8.16	5.95	-2.21	-27
1.0m	0.52	1.44	+0.92	+177

ENGINEERING STUDIES

of increased sand flow across the breakwater in addition to diffraction around the N.E. facing arm.

40. The only area of erosion as predicted is downdrift of Location 2. Recent bait digging over a large portion of this area by local fishermen has by disturbance probably exacerbated the situation. It is anticipated that as the area between Locations 1 and 2 fill and the accretion updrift of Location 3 develops this area will fill. The condition of this area, including the area leeward of Location 2 is being observed closely for any adverse developments. Beach levels immediately against the sea wall in this area actually show a marginal increase, thus the erosion trend has been reversed and the primary object of securing the wall's foundations has been achieved.

41. Between Location 3 and Location 4 shore connected accretion has been impressive and is estimated to be in the order of $400,000m^3$. Location 3 breakwater constructed across an old shipping channel interrupts much of the flow coming from the W and NW. Percolation through and over the breakwater, and diffraction at its seaward end have resulted in the formation of this extensive sand bank which continues to build and now covers the lower part of the sandstone rock outcrop. Location 4 shore-connected breakwater controls development of the bank eastwards.

STRUCTURAL PERFORMANCE

42. Only two types of damage have been observed to the rock armour at the Leasowe Bay breakwaters. Individual pieces of armour were dislodged from the crest of both breakwaters during severe storms in 1984. At the eastern end of the downdrift breakwater there has been considerable armour movement on the backslope where the breakwater crest runs into the roundhead. This end of the breakwater has been observed to be subject to higher energy levels thought to be due to the orientation of the breakwater relative to incident wave direction. The dislodged armourstone was retrieved from the breakwater backslopes, during one weeks operation in 1988, using a tracked excavator which ran along the crest of the breakwater. The cost of this operation was less than £2,000.

43. The beach seaward of the updrift breakwater has suffered no erosion and there is no evidence of undermining of the breakwater's foundations. At the downdrift breakwater the higher energy levels at the eastern end and flow from the bay resulted in the formation of a large scour hole. This hole threatened to locally undermine the bedstone anti-scour apron. Two small fingers of bedstone, constructed across the scour hole perpendicular to the outer face of the apron, have pushed flow seaward and secured the apron's foundation.

44. At Kings Parade, no measurable movement of the armourstone, which forms the lower perimeter of the breakwaters has taken place since construction.

45. Immediately following construction of each of the five breakwaters detailed structural inspections were carried out

after any significant sea states (> Force 6 in the N-W quadrant). The only signs of reef block movement were at Location 3, where a general movement of units, in the direction of incident wave energy was recorded. This minor and uniform movement closed up the wider joints on the outside of the radius of the breakwater, which is curved in plan. Subsequently the breakwaters have been inspected as part of the Authority's quarterly maintenance inspections, with additional surveys being carried out after significant storm events (> Force 8 in the N-W quadrant). At present the only observed damage to reef blocks has occurred towards the eastern end of Location 2. Four units, on the upper terrace file, have suffered minor damage during storm action. It is thought that these units were lifted and rotated, the movement being restrained by anti-rotation ribs and the weight of adjacent units. The damaged areas were repaired by breaking back behind the steel reinforcement and reforming the concrete section using a polymer modified concrete. The most recent inspections show that no further damage has occurred to Reef Block units.

46. The use of slender section concrete armour units, that are reinforced with steel, requires that the utmost attention is paid to the specification of durable concrete. Levels of site supervision must also be appropriate to ensure that the most durable concrete units are produced.

47. The University of Liverpool are at present engaged on a research project to investigate and evaluate various methods for the non-destructive testing of concrete. As part of this project the research team are now monitoring a small sample of Reef Blocks (10 out of a total of 2250) on a regular basis. This monitoring consists firstly of measuring the electro-chemical potential of the steel in relation to a stable half cell electrode placed in contact with the concrete surface (ref. 8). Combining potential results with information obtained from a resistivity survey carried out using a 4-point Wenner probe (ref. 9), is thought to give a good indication of the actual rate of corrosion activity.

48. The durability performance of concrete is directly related to its permeability. The passivating alkali environment provided by the concrete cover may be disrupted by the ingress of chlorides. Once corrosion has been initiated, both oxygen and water are needed for continued activation. Hence reinforced concrete that is permanently submerged or permanently dry is at the lowest risk. The results of the monitoring programme so far indicate that the durability performance of the Reef Blocks is good (ref. 10). It is thought that Reef Blocks represent a fairly low durability risk, due to their level in relation to tide levels limiting oxygen availability.

CONCLUSIONS

49. (a) the recorded trends in beach development are generally as predicted and indicate that the erosion trend has been halted

- (b) the structural performance of both rock armour and concrete units to date has been good, although modifications would be made for any future schemes
- (c) an appropriate level of research, design input and feasibility planning is required
- (d) forward planning, regarding material supplies and access can lead to cost savings and rapid construction
- (e) interface details, between different materials and the structure and beach need careful attention
- (f) breakwater schemes allow low cost fine tuning either during or after construction
- (g) navigation aspects should be investigated
- (h) the safety aspects for beach users and small boat users need to be considered at the design stage
- (i) several areas of beach now stay dry during neap tides, these have become focal points for day trippers. The Authority is now developing a management plan for access and use of the coastal strip.
- (j) the breakwaters have been colonised by a multitude of marine organisms
- (k) the creation of high beaches has increased the deposition of unsightly rubbish on the tide line

The benefits in terms of improved hydraulic performance, local beach development, local amenity intrusion and cost, are considered to outweigh any risks associated with the use of slender reinforced concrete units in the marine conditions on Wirral. This assessment is based on the trends established from the programme of performance monitoring.

Finally in this particular location of high energy levels and a strong littoral drift, the use of breakwaters in place of more conventional methods of coastline control, is considered to be an appropriate method of coastline management.

ACKNOWLEDGEMENTS

50. The Author is grateful for the permission of, and assistance from, Mr R L Bott, CEng.FICE.FIHT. Borough Engineer, Metropolitan Borough of Wirral, in the preparation of this paper. He is also indebted to his colleagues at the Metropolitan Borough of Wirral for the information and help that they provided. Finally he would like to thank Dr P C Barber MEng.PhD.MICE.MBIM (formerly Principal Engineer for Coast Protection at Wirral until the end of 1984) now Managing Director of CEEMAID, for his comments on the paper.

REFERENCES

1. BARBER P.C. and DAVIES C.D. Offshore Breakwaters - Leasowe Bay. Proceedings of the Institution of Civil Engineers, Part 1, 1985, Vol. 77, Feb. 85-109.
2. BARBER P.C. and LLOYD T.C. The diode wave absorption block. Proceedings of the Institution of Civil Engineers. Part 1, 1984, vol. 76, Nov. 847-870.
3. BARBER P.C. A preliminary investigation into the causes of beach erosion at Kings Parade on the North Wirral coast. Liverpool University, MEng thesis, 1977.

4. BARBER P.C. et al. A physical model study of the effects of an offshore breakwater at Wallasey Embankment. Metropolitan Borough of Wirral, Dept. of Engineering Services, Internal Report, 1978.
5. BARBER P.C. et al. Kings Parade field data investigation. Metropolitan Borough of Wirral, Dept. of Engineering Services. Internal Report, 1981, I-IV.
6. DAVIES C.D. Offshore Breakwaters at Wirral. Municipal Engineer, 1985, vol. 2, Aug. 198-207.
7. BURROWS R. et al. Wind/Wave/Analysis and Storm Simulation. Liverpool Bay, 1985, vol. I and II, March.
8. ASTM C 876-80. - ½ cell testing.
9. WENNER F. "A method for measuring earth resistivity". Bulletin of the Bureau of Standards. 1915. vol. 12.
10. MILLARD S.G. et al. Performance of Slender Reinforced Concrete Coastal Armour Units. Paper submitted to Coastal Engineering. Summer 1988.